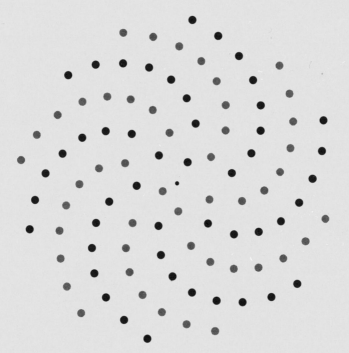

$$\sqrt{2}$$

The Square Root of 2

A Dialogue Concerning a Number and a Sequence

David Flannery

C

COPERNICUS BOOKS

An Imprint of Springer Science+Business Media

PRAXIS

in Association with

Praxis Publishing, Ltd.

Published in the United States by Copernicus Books,
an imprint of Springer Science+Business Media.

Copernicus Books
Springer Science+Business Media
233 Spring Street
New York, NY 10013
www.springeronline.com

Library of Congress Control Number:
2005923268

Manufactured in the United States of America.
Printed on acid-free paper.

9 8 7 6 5 4 3 2 1

ISBN-10: 0-387-20220-X
ISBN-13: 978-0387-20220-4

Why, sir, if you are to have but one book with you on a journey, let it be a book of science. When you have read through a book of entertainment, you know it, and it can do no more for you; but a book of science is inexhaustible. . . .

—James Boswell
Journal of a Tour to the Hebrides with Samuel Johnson

Contents

Prologue

You may think of the dialogue you are about to read, as I often did while writing it, as being between a "master" and a "pupil"—the master in his middle years, well-versed in mathematics and as devoted and passionate about his craft as any artist is about his art; the pupil on the threshold of adulthood, articulate in speech, adventuresome of mind, and enthusiastically receptive to any knowledge the more learned teacher may care to impart.

Their conversation—the exact circumstances of which are never described—is initiated by the master, one of whose tasks is to persuade his disciple that the concept of number is more subtle than might first be imagined. Their mathematical journey starts with the teacher guiding the student, by way of questions and answers, through a beautifully simple geometrical demonstration (believed to have originated in ancient India), which establishes the existence of a certain number, the understanding of whose nature is destined to form a major part of the subsequent discussion between the enquiring duo.

Strong as the master's motivation is to have the younger person glimpse a little of the wonder of mathematics, stronger still is his desire to see that his protégé gradually becomes more and more adept at mathematical reasoning so that he may experience the pure pleasure to be had from simply "finding things out" for himself. This joy of discovery is soon felt by the young learner, who having embarked upon an exploration, is richly rewarded when, after some effort, he chances upon a sequence of numbers that he surmises is inextricably linked to the mysterious number lately revealed by the master. Enthralled by this fortunate occurrence, he immediately finds himself in the grip of a burning curiosity to know more about this number and its connection with the sequence that has already captivated him. Thus begins this tale told over five chapters.

I have made every effort to have the first four chapters as self-contained as possible. The use of mathematical notation is avoided whenever words can achieve the same purpose, albeit in a more lengthy manner. When mathematical notation is used, nothing beyond high school algebra of the simplest kind is called on, but in ways that show clearly the need for this branch of mathematics. While the algebra used is simple, it is often clever, revealing that a few tools handled with skill can achieve a great deal. If readers were to appreciate nothing more than this aspect of algebra—its power to prove things in general—then this work will not have been in vain.

Unfortunately, to have the fifth chapter completely self-contained would have meant sacrificing exciting material, something I didn't wish to do, preferring to reward the reader for the effort taken to reach this point, when it is hoped he will understand enough to appreciate the substance of what is being related.

Throughout the dialogue, so as to distinguish between the two speakers, the following typographical conventions are used:

The Master's Voice—assured, but gently persuasive—is set in this mildly bold typeface, and is firmly fixed at the left edge of the column.

The Pupil's Voice—deferential, but eager and inquiring—is set in this lighter font, and is moved slightly inward from the margin.

The best conversations between teachers and students are both serious and playful, and my hope is that the readers of this book will sense that something of that spirit, of real learning coupled with real pleasure, coexist in this dialogue.

David Flannery
September, 2005

$$\sqrt{2}$$

CHAPTER 1

Asking the Right Questions

I'd like you to draw a square made from four unit squares.

A unit square is one where each of the sides is one unit long?

Yes.

Well, that shouldn't be too hard.

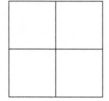

Will this do?

Perfect. Now let me add the following diagonals to your drawing.

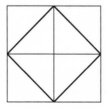

You see that by doing this a new square is formed.

I do. One that uses a diagonal of each of the unit squares for its four sides.

Let's shade this square and call it the "internal" square.

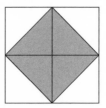

Now, I want you to tell me the area of this internal square.

> Let me think. The internal square contains exactly half of each unit square and so must have half the area of the large square. So it has an area of 2 square units.

Exactly. Now, what is the length of any one of those diagonals that forms a side of the internal square?

> Off-hand I don't think I can say. I know that to get the area of a rectangular region you multiply its length by its breadth.

"Length by breadth," as you say, meaning multiply the length of one side by the length of a side at right angles to it.

> So, for a square, this means that you multiply the length of one side by itself, since length and breadth are equal.

Yes.

> But where does this get me? As I said, I don't know the length of the side.

As you say. But if we let s stand for the length of one of the sides, then what could you say about s?

> I suppose there is no way that we could have this little chat without bringing letters into it?

There is, but at the cost of the discussion being more longwinded than it need be. Incidentally, why did I chose the letter s?

> Because it is the initial of the word side?

Precisely. It is very common to use the initial of the word describing the quantity you're looking for.

> So s stands for the length of the side of the internal square. I hope you are not going make me do algebra.

Just a very small amount—for the moment. So can you tell me something about the number s?

> When you multiply s by itself you get 2.

Exactly, because the area of the internal square is 2 (squared units). Do you recall that $s \times s$ is often written as s^2?

I do. My algebra isn't *that* rusty.

So you are saying that the number s "satisfies" the equation:

$$s^2 = 2$$

In words, "s squared equals two."

Okay, so the number s when multiplied by itself gives 2. Doesn't this mean that s is called the square root of 2?

Well, it would be more accurate to say that s is *a* square root of 2. A number is said to be a square root of another if, when multiplied by itself, it gives the other number.

So 3 is a square root of 9 because $3 \times 3 = 9$.

As is -3, because $-3 \times -3 = 9$ also.

But most people would say that the square root of 9 is 3.

True. It is customary to call the positive square root of a number its square root. And since s is the length of the side of a square, it is obviously a positive quantity, so we may say . . .

. . . that s is the square root of 2.

Sometimes, we simply say "root two," it being understood that it's a square root that is involved.

And not some other root like a cube root?

Yes. Now the fact that 3 is the square root of 9 is often expressed mathematically by writing $\sqrt{9} = 3$.

I've always liked this symbol for the square root.

It was first used by a certain Christoff Rudolff in 1525, in the book *Die Coss*, but I won't go into the reasons why he chose it.

Can we say goodbye to s and write $\sqrt{2}$ in its place from now on?   [See chapter note 1.]

If we want to, but we'll still use s if it serves our purposes.

So we have shown that the diagonal of a unit square is $\sqrt{2}$ in length.

Indeed we have. This wonderful way of establishing the existence of the square root of 2 originated in India thousands of years ago.   [See chapter note 2.]

You'd have to say that it is quite simple.

Which makes it all the more impressive.

So what number is $\sqrt{2}$?

As the equation $s^2 = 2$ says, it is the number that, when multiplied by itself, gives 2 exactly. This means no more or no less than what the equation

$$\sqrt{2} \times \sqrt{2} = 2$$

says it means: $\sqrt{2}$ is the number that when multiplied by itself gives 2.

I know, but what number does $\sqrt{2}$ actually stand for? I mean $\sqrt{16} = 4$, and 4 is what I would call a tangible number.

I understand. You have given me a concrete value for $\sqrt{16}$, namely the number 4. You want me to do the same for $\sqrt{2}$, that is, to show you some number of a type with which you are familiar, and that when squared, gives 2.

Exactly. I'm simply asking what the concrete value of s is, that makes $s^2 = 2$.

I can convince you quite easily that $\sqrt{2}$ is not a natural number.

The natural numbers are the ordinary counting numbers, 1,2,3, and so on.

Precisely.

Even though 2 itself is a natural number? The natural numbers 9 and 16 have square roots that are also natural numbers.

That's true, they do.

But you are saying that 2 doesn't.

I am. One way of seeing this is to write the first few natural numbers in order of increasing magnitude in a line, and beneath them on a second line write their corresponding squares:

$$1 \quad 2 \quad 3 \quad 4 \quad 5 \quad 6 \quad 7 \ldots$$
$$1 \quad 4 \quad 9 \quad 16 \quad 25 \quad 36 \quad 49 \ldots$$

The three dots, or ellipsis, at the end of a line means that the pattern continues without stopping.

Well, I can see straight away that the number 2 is missing from the second row.

As are

$$3, \quad 5, \quad 6, \quad 7, \quad 8, \quad 10, \quad 11, \quad 12, \quad 13, \quad 14, \quad 15, \quad 17, \ldots$$

I would say that there are a lot more numbers missing than are present.

Yes, in a sense "most" of the natural numbers are absent from this second line. The numbers 1, 4, 9, 16, . . . that appear on it are known as the *perfect squares*.

And those numbers that are missing from this line are not perfect squares?

Correct: 49 is a perfect square but 48 is not.

I think I see now why there is no natural number squaring to 2. The first natural number squares to 1 while the second natural number squared is 4, so 2 gets skipped over.

That's about it.

All right. It is fairly obvious, now at any rate, that there is no natural number that squares to 2, but surely there is some fraction whose square is 2?

By fraction, you mean a common fraction where one whole number is divided by another whole number?

That's what I mean, $\frac{7}{5}$, for example. Are there other types of fractions?

There are, but when we say "fraction" we mean one whole number divided by another one. The number being divided is the numerator and the one doing the dividing is called the denominator.

The number on the top is the numerator and the number on the bottom is the denominator.

That's it exactly. In your example, the whole number 7 is the numerator while the whole number 5 is the denominator.

Now mustn't there be some fraction close to this one that squares to give 2 exactly?

Why did you say *close* to this one?

Because my calculator tells me that $\frac{7}{5}$ is 1.4 in decimal form; and when I multiply this by itself I get 1.96, which is fairly close to 2.

Agreed. Let me show you how we can see this for ourselves without a calculator but using a little ingenuity instead. Since

$$\left(\frac{7}{5}\right)^2 = \frac{49}{25}$$
$$\overset{!}{=} \frac{50-1}{25}$$
$$= \frac{50}{25} - \frac{1}{25}$$
$$= 2 - \frac{1}{25}$$

we can say that the fraction $\frac{7}{5}$ when squared underestimates 2 by the amount $\frac{1}{25}$.

And according to my calculator $\frac{1}{25} = 0.04$, which is just $2 - 1.96$. By the way, why did you put the exclamation point over the second equals sign?

To indicate that the step being taken is quite a clever one.

It certainly wouldn't have occurred to me, which I know is not saying much.

Well, I don't lay any claim to originality for taking this step. I have seen many similar such tricks used by others in the past and, after all, I knew what it was I wanted to show.

At least I can see why it's clever.

Good. Why?

By writing the numerator 49 as 50 − 1, you were able to divide the 50 by 25 to get 2 exactly and the 1 by 25 to get $\frac{1}{25}$ as the measure of the underestimate.

Desert Island Math

A useful trick if you're stranded on a desert island without any calculating devices other than your own poor head.

Pure do-it-yourself mathematics! I suppose using a calculator to get the value of something you wouldn't be able to calculate for yourself is a form of cheating?

Do you mean like asking for the decimal expansion of $\sqrt{2}$, for example?

Well, something like that. I wouldn't have a clue how to get the decimal expansion of $\sqrt{2}$ using my own very limited powers.

$$\begin{array}{r} 1.4 \\ 5\overline{\smash)7.00} \\ \underline{5} \\ 20 \\ \underline{20} \\ 00 \end{array}$$

I'm sure you do your mental abilities an injustice. If we know and understand how to get a decimal expansion of a number "by hand," then we don't contravene the DIY philosophy if we use a calculator to save labor.

Are you saying that because I know how to get the decimal expansion of $\frac{7}{5}$ or $\frac{3}{11}$ by long division, even though I wouldn't like to be pressed on why the procedure works, I may use a calculator to avoid the "donkey work" involved with such a task?

algorithm: step-by-step procedure

I think we'll let this be a policy. We'll assume that if we were put to it we could explain to ourselves and others the "ins and outs" of the long-division algorithm.

Of course, completely!

$$\begin{array}{r} 0.272\ldots \\ 11\overline{\smash)3.000} \\ \underline{22} \\ 80 \\ \underline{77} \\ 30 \\ \underline{22} \\ 80 \\ \vdots \end{array}$$

Decimal expansions, or "decimals" as we often say for short, have certain advantages, one being that they convey the magnitude of a number more readily than their equivalent fractions do. When a number is expressed in decimal form, it is easy to say geometrically where it is located on the number line. No matter how long the decimal expansion of a number may be, we still know between which two whole numbers it lies on this number line:

So we can see quite easily from 1.4 that it is a number between 1 and 2, whereas it is not as easy to see this from the fraction $\frac{7}{5}$.

The fraction $\frac{7}{5}$ is perhaps too simple. It is not too difficult to mentally determine the two whole numbers between which it is located on the number line, but who can say without resorting to a calculation where the fraction $\frac{103993}{33102}$ is positioned on the same line?

I see the point, or should I say I do not see the (decimal) point!

Hmm! Speaking of the fraction $\frac{7}{5}$, you might like to get a box of matches and construct a square with five matches on each side.

Does this mean that the five matches between them make up the unit-length?

You can certainly think of it this way, if you like. Now you'll find that seven matches will fit along the diagonal:

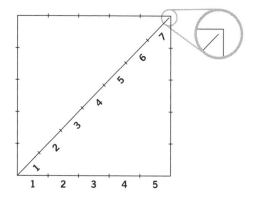

These seven matches do not stretch the full length of the diagonal since $\frac{7}{5}$ underestimates $\sqrt{2}$.

That they don't is barely visible.

True, but the gap is there.

This is a rather neat way of visualising $\frac{7}{5}$ as an approximation to $\sqrt{2}$.

Yes it is, isn't it? Looked at another way it says that the ratio $7:5$ is close to the ratio $\sqrt{2}:1$. Now, where were we?

Looking for a fraction that squares to 2.

Indeed, so let's continue the quest. Any further thoughts?

There must be some fraction a little bit bigger than $\frac{7}{5}$ that squares to give 2 exactly.

Well, there are lots of fractions just a little bit bigger than $\frac{7}{5}$.

I know. Isn't there an infinity of fractions between 1.4 and 1.5 alone?

Yes, but that this is so we can leave for another time. Why do you mention 1.5?

Simply because $(1.4)^2 = 1.96$ is less than 2 while $(1.5)^2 = 2.25$ is greater than 2.

So?

Doesn't this mean that the square root of 2 lies between these two values?

It does. In fact since $1.5 = \frac{3}{2}$ we may write that

The symbol < means "less than."

$$\frac{7}{5} < \sqrt{2} < \frac{3}{2}$$

Let me display this arithmetic "inequality" on the number line:

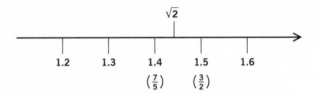

Notice that I have placed $\sqrt{2}$ to the right of 1.4 and closer to 1.4 than to 1.5 because $\frac{3}{2}$ squared overestimates 2 by $\frac{1}{4}$, which is much more than the $\frac{1}{25}$ by which $\frac{7}{5}$ squared underestimates $\sqrt{2}$.

But how do you locate $\sqrt{2}$ on the number line if you don't know what fraction it is?

A good question. The answer is that you do so geometrically.

I'd like to see how.

It's easy to construct a unit square geometrically on the interval that stretches between 0 and 1:

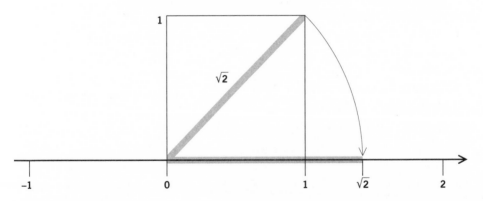

Now imagine the diagonal with one end at 0 and of length $\sqrt{2}$ being rotated clockwise about the point 0 until its other end lies on the number line.

At a point $\sqrt{2}$ from 0. Very smart.

Of course, this is an ideal construction where everything can be done to perfection.

I understand. It is the method that counts.

Yes.

An Exploration

But to return to the point I was making: surely among the infinity of fractions lying between 1.4 and 1.5 there is one that squares to give 2 exactly.

Well if there is, how do you propose finding it?

That's what is bothering me.

I'm sure you'll agree that it's not wise to begin checking fraction after fraction in this infinity of fractions without having some kind of plan.

Absolutely, it could take forever. What would you suggest?

Thinking about the problem a little to see if we can find some systematic way of attacking it.

Sounds as if we are about to go into battle.

A mental battle. Let us begin our campaign by examining the implications of expressing the number $\sqrt{2}$ as a fraction.

This could get interesting. What are you going to call this fraction?

Well, since we don't know it, at least not yet, we must keep our options open. One way of doing this is to use distinct letters, one to stand for its numerator and the other for its denominator.

Here comes some more algebra.

Only a little, used as scaffolding as it were, just to get us started.

Well, I'll stop you if I think I'm losing the drift of the discussion.

Let's call the numerator of the fraction m and the denominator n.

So if the fraction were $\frac{7}{5}$, which I know it is not, then m would be 7 and n would equal 5.

Or put slightly differently, if $m = 7$ and $n = 5$ then

$$\frac{m}{n} = \frac{7}{5}$$

I'm with you.

Now if

$$\sqrt{2} = \frac{m}{n}$$

then

$$\sqrt{2} \times \sqrt{2} = \frac{m}{n} \times \frac{m}{n}$$

Agreed?

I think so. You are simply squaring both sides of the original equation.

I am, and I do so in this elaborate manner to highlight the presence of $\sqrt{2} \times \sqrt{2}$.

Which by definition is 2.

Yes, a simple but vital use of the defining property of $\sqrt{2}$, which allows us to write that

$$2 = \left(\frac{m}{n}\right)^2$$

We can turn this equation around and write

$$\left(\frac{m}{n}\right)^2 = 2$$

to put the emphasis on the fraction $\frac{m}{n}$. What is the equation saying about $\frac{m}{n}$?

That its square is 2.

Exactly. And since

$$\left(\frac{m}{n}\right)^2 = \frac{m^2}{n^2}$$

we can say that

$$\frac{m^2}{n^2} = 2$$

or that

$$m^2 = 2n^2$$

So this equation is a consequence of writing $\sqrt{2}$ as $\frac{m}{n}$?

It is indeed. Now let us see what we can learn from it.

I'll leave this to you.

I'm sure it won't be long before you join in. For one thing, $m^2 = 2n^2$ tells us is that if we are to find a fraction that is equal to $\sqrt{2}$, then we must find two perfect squares, one of which is twice the other.

What are perfect squares again? Oh, I remember, 1, 4, 9, 16, . . . That's right, a perfect number is one that is the square of a natural number.

Well, this is a task that I can definitely undertake.

Be my guest.

Why don't I make out a list of the first twenty squares along with their doubles and see if I can find a match between some square and the double of some other square.

An excellent plan. Nothing like a bit of "number crunching," as it's called, to really get one thinking.

Of course, I'm going to use a calculator just to speed things up.

Naturally. Nobody doubts that you can multiply one number by itself.

Here's the table I get:

Natural Number	Number Squared	Twice Number Squared
1	1	2
2	4	8
3	9	18
4	16	32
5	25	50
6	36	72
7	49	98
8	64	128
9	81	162
10	100	200
11	121	242
12	144	288
13	169	338
14	196	392
15	225	450
16	256	512
17	289	578
18	324	648
19	361	722
20	400	800

The three columns show, in turn, the first twenty natural numbers, their squares, and twice these squares.

Great. We can think of the second column as corresponding to m^2 numbers and the third column as corresponding to numbers of the form $2n^2$.

I'm not sure I understand what you are saying here.

I'll explain by example. We may think of the number 196 in the second column as being an m^2 number, where $m = 14$, while we may consider the number 450 in the third column as being a $2n^2$ number, where $n = 15$.

Let me test myself to see if I have got the idea. I can think of 16 in the second column as an m^2 number with $m = 4$, while I can think of the 648 in the third column as corresponding to $2n^2$, with $n = 18$, because $2(18)^2 = 648$. Do I pass?

With honors. Now if you can find an entry in the second column that matches an entry in the third column, you will have found values for m and an n which make $m^2 = 2n^2$ and so you'll have a fraction $\frac{m}{n}$ equal to $\sqrt{2}$.

As easy as that? So fingers crossed as I look at each entry of the second column of this table and then look upwards from its location along the third column for a possible match.

Of course! A time-saving observation. As you say, you need only look upwards because the corresponding entries in the third column are bigger than those in the second.

Unfortunately, I can't find a single entry in the second column that is equal to any entry in the third column.

So the second and third columns have no element in common.

Not that I can see. I'm going to experiment a little more by calculating the next ten perfect squares along with their doubles.

Good for you.

This time I get:

Natural Number	Number Squared	Twice Number Squared
21	441	882
22	484	968
23	529	1058
24	576	1152
25	625	1250
26	676	1352
27	729	1458

Natural Number	Number Squared	Twice Number Squared
28	784	1568
29	841	1682
30	900	1800

I realize that this is not much of an extension to the previous table.

Maybe, but perhaps you'll get a match this time.

I'm scanning the second column to see if any entry matches anything in the previous third column or the new third column.

Any luck?

I'm afraid not. However, I notice that there are some near misses in the first table.

What do you mean by "near misses"?

A discovery?

There are entries in the second column that are either just 1 less or 1 more than an entry in the third column.

I'm more than curious; please elaborate.

Well, take the number 9 in the second column. It is 1 more than the 8 in the third column.

True. Any others?

There's a 49 in the second column that is 1 less than the 50 appearing in the third column.

Again, true. Any more?

Yes. There's a 289 in the second column and a 288 in the third column.

Again, as you observed, with a difference of 1 between them. Did you find any more examples?

Not that I can see in these two tables, except, of course, at the very beginning. There's a 1 in the second column and a 2 in the third column.

Indeed there is.

But I don't know what to make of these near misses.

However, you seem to have hit upon something interesting, exciting even, so let's take a little time out to mull over your observations.

Fine by me, but you'll have to do the thinking.

Why don't we look at the case of the 9 in the second column and the 8 in the third column. What is the m number corresponding to this 9 in the second column, and what is the n number corresponding to the 8 in the third column?

Let me think. I would say that $m = 3$ and that $n = 2$.

And you'd be right. Your observation tells us that

$$m^2 = 2n^2 + 1$$

where $m = 3$ and $n = 2$.

> Because $3^2 = 2(2)^2 + 1$?

Exactly. Now let us move on to the case of the number 49 in the second column and the 50 in the third column.

> Here the $m = 7$ and $n = 5$ since $2(5)^2 = 2(25) = 50$.

This time

$$m^2 = 2n^2 - 1$$

where $m = 7$ and $n = 5$.

> Can I try the next case?

By all means.

> The number 289 corresponds to $m = 17$ since in this case $m^2 = 289$. On the other hand, the number 288 corresponds to $n = 12$ since $2(12)^2 = 2(144) = 288$.

No argument there.

> This time

$$m^2 = 2n^2 + 1$$

where $m = 17$ and $n = 12$.

So we're back to 1 over.

> There seems to be an alternating pattern with these pairs of near misses.

There does indeed. For the sake of completeness, you should look at the first case.

> You mean the case with 1 in the second column and 2 in the third column?

None other; the smallest case, so to speak.

> Okay. Here $m = 1$ and $n = 1$.

And what is the value of $m^2 - 2n^2$ on this occasion?

> This time

$$m^2 = 2n^2 - 1$$

Does this fit the alternating pattern?

> It does.

Which is great.

> But returning to the original reason for constructing the tables, I haven't found a single square among the first thirty perfect squares that is equal to twice another square.

True, and that means that, as of yet, you haven't found a fraction $\frac{m}{n}$ that squares to 2. But, on the other hand, you have found a number of very interesting fractions.

> I have? I would have thought that I've only found pairs of natural numbers that are within 1 of each other.

In a sense, you could say that. But you actually have discovered fractions with the property that the square of their numerator is within 1 of double the square of their denominator.

> I'm afraid you'll have to elaborate.

Of course. You remember we said, when you observed that 9 in the second column was 1 greater than the 8 in the third column, that the 9 corresponded to m^2 where $m = 3$, while the 8 corresponded to $2n^2$ where $n = 2$?

> I do.

Furthermore, $m^2 - 2n^2 = 1$, in this case.

> That's correct.

Suppose now that we form the fraction

$$\frac{m}{n} = \frac{3}{2}$$

Then can't we say that the equation $m^2 - 2n^2 = 1$ tells us that this fraction is such that the square of its numerator is 1 more than twice the square of its denominator?

> It seems to. I'll have to think a little more about this. Yes: $3^2 = 9$ and $2(2)^2 = 8$.

Try another one. Ask yourself, "What fraction is associated with the observation that the 49 in the second column is 1 less than the 50 in the third column?"

> Okay. Here $m = 7$ and $n = 5$, so the fraction is $\frac{7}{5}$, right?

Absolutely. Now what can you say about the numerator and denominator of this fraction?

> That the square of the numerator is 1 less than twice the square of its denominator.

Exactly.

> I think I understand now. You are saying that every time we observe the near miss phenomenon we actually find a special fraction.

Yes. You looked for a fraction whose numerator squared would match twice its denominator squared; you didn't find one, but instead you found fractions whose numerators squared are within 1 of twice their denominators squared.

> That's a nice way of looking at it.

Often when you look for something specific you chance upon something else.

> So I suppose you could say that I found the next best thing.

I think we can say this, and not a bad reward for your labors.

> Actually, I'm really curious to know if there are any more than just these four misses and to see if the plus or minus pattern continues to hold.

Let's hope so. Why don't we do a little more exploring?

> I'd be happy to but shouldn't we stick to our original mission of finding a difference of exactly 0?

Very nicely put. Finding an m and n such that $m^2 = 2n^2$ means that the difference $m^2 - 2n^2$ would be 0.

> Thanks.

However, I think we'll indulge ourselves and investigate your observation about near misses a little further, particularly as it looks so promising.

> Okay. I'll extend my tables and then go searching.

You could do that, but it might be an idea to look more carefully at what you have already found.

> Like good scientists would.

As you say. Begin by cataloguing the specimens found to date and examine them carefully for any clues.

> Will do.

Time to Reflect

> Beginning with the smallest, and listing them in increasing order, the fractions are

$$\frac{1}{1}, \quad \frac{3}{2}, \quad \frac{7}{5}, \quad \frac{17}{12}$$

Not many as of yet, but tantalizing.

> What secrets do they hold, if any?

Indeed. Can you spot some connection between them?

> Just like one of those sequence puzzles, "What is the next number in the sequence?" except here it looks harder because these are fractions and not ordinary numbers.

A puzzle certainly, but one we have encountered quite naturally.

> And not just made up for the sake of it.

Yes, something like that.

I hope this is an easy puzzle.

It is always best to be optimistic so I advise that you say to your-self, "This is sure to be easy," and look for simple connections.

Optimism it is then, but where to start?

It is often a good idea to begin by examining a pair of terms some way out along a sequence rather than at the very begin-ning of it.

Right. On that advice I'll see if I can spot a connection between

$$\frac{7}{5} \quad \text{and} \quad \frac{17}{12}$$

and if I think I have found one, I'll check it on the earlier fractions.

Very sensible. Happy hunting!

I think I'll begin by focusing on the denominator 12 of the fraction $\frac{17}{12}$.

Following a very definite line of inquiry, as they say.

I think I have spotted something already.

Which is?

That $12 = 7 + 5$, the next denominator looks as if it might be the sum of the numerator and denominator of the previous fraction.

If it's true, it will be a big breakthrough. I must say that was pretty quick.

I must check the earlier terms of the sequence

$$\frac{1}{1}, \quad \frac{3}{2}, \quad \frac{7}{5}, \quad \frac{17}{12}$$

to see if this rule holds also for their denominators.

Fingers crossed, then.

I obviously cannot check the first fraction, $\frac{1}{1}$.

Why not?

Because there is no fraction before it.

A good point.

But the second fraction, $\frac{3}{2}$, has denominator 2, which is just $1 + 1$, the sum of the numerator 1 and denominator 1 of the first fraction $\frac{1}{1}$. This is getting exciting.

That's great. How about the third fraction $\frac{7}{5}$?

> Right, Mr. $\frac{7}{5}$, let's see if you fit the theory. Your denominator is 5, is it not? Indeed it is, and the sum of the numerator and denominator of the previous fraction, $\frac{3}{2}$, is $3 + 2$, which I'm happy to say is none other than 5. This is fantastic! Who would have thought?

Great again! Now is there an equally simple rule for the numerators?

> I hope so, because discovering that rule for the denominators gave me a great thrill.

We couldn't ask for more than that.

> Right, back to the drawing board. So is there a connection between the numerator 17 of the fraction $\frac{17}{12}$ and the numbers 7 and 5 from the previous fraction $\frac{7}{5}$?

It would be marvelous if there were.

> If I'm not mistaken, there is. It's simply that $17 = 7 + (2 \times 5)$.

Well spotted, though not quite as simple as the rule for the denominators.

> No, but still easy enough.

Once you see it. How do you interpret this rule?

> Doesn't it say that the next numerator is obtained by adding the numerator of the previous fraction to twice the denominator of the previous fraction?

Indisputable. You had better check this rule on the other fractions.

> It works for the fraction $\frac{3}{2}$ since $3 = 1 + (2 \times 1)$, and it also works for $\frac{7}{5}$ since $7 = 3 + (2 \times 2)$.

This is wonderful. So how would you summarize the overall rule, which allows one to go from one fraction to the next?

> Well, the general rule obtained by combining the denominator rule and the numerator rule seems to be:

>> To get the denominator of the next fraction, add the numerator and denominator of the previous fraction; to get the numerator of the next fraction, add the numerator of the previous fraction to twice its denominator.

Well done! And a fairly straightforward rule, at that.

> Isn't it amazing?

Indeed it is. After all, there was no reason to believe that there had to be any rule whatever connecting the fractions, but to find that there is one and that it's simple is remarkable.

> I must now apply this general rule to $\frac{17}{12}$ to see what fraction comes out and to see if it has the property that its top squared minus twice the bottom squared is either 1 or -1.

Let's hope that the property holds.

> According to the rule, the next fraction has a denominator of $17 + 12 = 29$ and a numerator of $17 + 2 \times 12 = 41$, and so is $\frac{41}{29}$.

Good. And now what are we hoping for?

> Based on the pattern displayed by the previous four fractions, that $(41)^2 - 2(29)^2$ will work out to be -1.

Perform the acid test.

> Here goes:
>
> $$41^2 - 2(29)^2 = 1681 - 2(841) = 1681 - 1682 = -1$$
>
> This is fantastic!

So now you have found another example of a perfect square that is within 1 of twice another perfect square—the whole point of this investigative detour—*without* having to go to the bother of extending your original tables.

> That's true. Our more thorough examination of the four cases we found seems to have paid off.

A little thought can save a lot of computing.

> I know that I couldn't have spotted this example with my tables because they give only the first thirty perfect squares along with their doubles; but can we be sure that there is not an m value between 17 and 41 that gives a square that is within 1 of twice another perfect square?

An excellent question. At the moment we can't be sure without checking. However, if there is such an m value, then it corresponds to a fraction $\frac{m}{n}$ that doesn't fit in with the above rule. Of course, this doesn't exclude the possibility of there being such a value. However, if you check, you won't find any such value.

> I must calculate the next fraction generated by the rule to see if it also satisfies the plus or minus 1 property, to give it a name. Applying the rule to $\frac{41}{29}$ gives $41 + 29 = 77$ as the next denominator and $47 + (2 \times 29) = 99$ as the numerator.

So $\frac{99}{70}$ is the next fraction to be tested.

> I predict that $m^2 - 2n^2 = 1$ in this case. The calculation
>
> $$99^2 - 2(70)^2 = 9801 - 2(4900) = 9801 - 9800 = 1$$
>
> verifies this. Great!

Bravo! What now?

> Obviously, we can apply the rule over and over again and so generate an infinite sequence beginning with
>
> $$\frac{1}{1}, \frac{3}{2}, \frac{7}{5}, \frac{17}{12}, \frac{41}{29}, \frac{99}{70}, \ldots$$

True, you can generate an infinite number of fractions using the rule but . . .

> How can we be sure that all the fractions of this sequence have the property that $m^2 - 2n^2$ is either plus or minus 1 without checking each, which is clearly out of the question.

Yes, this is a bit of a problem. It might be that answering such a question may prove difficult or even impossible.

> And can we say that these are the only fractions having this plus or minus 1 property?

My, my, what truly mathematical questions! You need have no fear that you and mathematics are strangers if you can think up questions like this.

> I don't know about that. Normally, I know I wouldn't even dream of asking questions such as these, but at the moment my mind seems to be totally engrossed by these particular fractions.

Ah, well, I've read somewhere that you really only see a person's true intelligence when his or her affections are fully engaged.

> Perhaps tomorrow I won't care, but right now I really want to know if all the fractions generated by the rule actually obey the plus or minus 1 property; and I also want to know if these are the only fractions that do so.

Good for you. In mathematics it often seems that asking questions is the easy part, whereas it is the answering of them that is hard. But asking the right questions is a very important part of any investigation, whether it be mathematical or otherwise.

> The good detectives always ask the right questions.

Well, by the end at any rate.

> But can you tell me if my questions have answers; and if they do, what are their answers?

They do, but I am not going to tell you what they are. I don't want to spoil the fun you'll have in trying to answer them for yourself later.

> Later could be an eternity away if it is left up to me on my own.

That remains to be seen. Anyway, you have opened up a rich vein for further exploration with your observation that there are squares whose doubles are within plus or minus 1 of other squares, and with your recent rule, both of which we'll come back to soon.

> So, are we going to return to our original investigation?

Not just yet.

Squeezing $\sqrt{2}$

Before leaving the fractions

$$\frac{1}{1}, \frac{3}{2}, \frac{7}{5}, \frac{17}{12}, \frac{41}{29}, \frac{99}{70}, \cdots$$

let me show you how they connect with the number $\sqrt{2}$ itself.

Although none of them is $\sqrt{2}$?

Correct. But each of them can be thought of as an *approximation* to $\sqrt{2}$. In fact, each successive fraction provides a better approximation to $\sqrt{2}$ than its predecessor.

I hope you don't mind my saying so, but I would be much more interested in finding the exact fraction instead of approximations, however good they might be.

I appreciate that you are impatient to get on with your searching, but follow me for just a little longer so that I can show you how simply but cleverly we can use these fractions to close in on the location of $\sqrt{2}$ on the number line.

All right. Maybe I'll learn something that will help with my search.

Perhaps; we should look for clues anywhere we can. Now we know that

$$1^2 = 2(1)^2 - 1$$
$$3^2 = 2(2)^2 + 1$$
$$7^2 = 2(5)^2 - 1$$
$$17^2 = 2(12)^2 + 1$$
$$41^2 = 2(29)^2 - 1$$
$$99^2 = 2(70)^2 + 1$$

Yes.

These equations are either of the form

$$m^2 = 2n^2 - 1$$

or

$$m^2 = 2n^2 + 1$$

alternating between one and the other.

Agreed. And I would bet that this jumping between -1 and 1 continues forever, though I have no idea how to prove it.

Now let us divide each of the equations

$$1^2 = 2(1)^2 - 1$$
$$3^2 = 2(2)^2 + 1$$
$$7^2 = 2(5)^2 - 1$$
$$17^2 = 2(12)^2 + 1$$
$$41^2 = 2(29)^2 - 1$$
$$99^2 = 2(70)^2 + 1$$

by their corresponding n^2 values to get

$$\left(\frac{1}{1}\right)^2 = 2 - \frac{1}{1^2}$$

$$\left(\frac{3}{2}\right)^2 = 2 + \frac{1}{2^2}$$

$$\left(\frac{7}{5}\right)^2 = 2 - \frac{1}{5^2}$$

$$\left(\frac{17}{12}\right)^2 = 2 + \frac{1}{12^2}$$

$$\left(\frac{41}{29}\right)^2 = 2 - \frac{1}{29^2}$$

$$\left(\frac{99}{70}\right)^2 = 2 + \frac{1}{70^2}$$

Are you with me?

Just about. I'm still mentally dividing across the second equation by 2^2, putting it beneath the 3^2 and placing the combination $\frac{3}{2}$ under one umbrella with the power of 2 outside.

Takes practice, but it's all legal.

I'll accept this, since you did it, but I'm a little rusty when it comes to fractions and powers, so I can be slow. Anyway, I'm happy with this last set of equations now.

This simple but clever idea gives us equations that are very informative. They tell, in turn, how close the square of each fraction is to the number 2. Can you see why?

I'll have to take time on this. What is the equation

$$\left(\frac{17}{12}\right)^2 = 2 + \frac{1}{12^2}$$

saying? That when we square $\frac{17}{12}$ we get 2 plus the fraction $\frac{1}{144}$?

Yes. And?

Since $\frac{1}{144}$ is small, the fraction $\frac{17}{12}$ isn't a bad approximation of $\sqrt{2}$.

Not bad at all.

I think I see now why the approximations are getting better and better. As we move down through the set of equations, the fractions on the very right-hand side get smaller and smaller.

True. So?

So the corresponding fractions squared on the right-hand side are getting closer and closer to 2, which I take it means that the fractions themselves are better and better approximations of $\sqrt{2}$.

Excellent. We can say more.

We can?

We can say that the alternate fractions

$$\frac{1}{1}, \quad \frac{7}{5}, \quad \frac{41}{29}$$

are three underestimates of $\sqrt{2}$, each being a better approximation of $\sqrt{2}$ than its predecessor.

Because the minus sign before the last fraction in each equation tells us that these fractions squared are less than 2 by some amount.

That's it. The fraction $\frac{1}{1}$ is the smallest of these fractions, and $\frac{41}{29}$ is the largest:

You'll understand why I make this point in a moment.

But the fractions we skipped

$$\frac{3}{2}, \quad \frac{17}{12}, \quad \frac{99}{70}$$

on the other hand, are three overestimates of $\sqrt{2}$, which become progressively closer to $\sqrt{2}$.

Right again. When these fractions are squared, they give 2 plus something positive. Note that this time the first fraction is the largest and the last one the smallest.

This is the opposite of the previous case.

I think I see what you're driving at. The underestimates are creeping up on $\sqrt{2}$ from the left while the overestimates are creeping back toward $\sqrt{2}$ from the right.

That's right, as we can see when we show all six fractions together on the number line:

Here is one way of summarizing this information:

$$\frac{1}{1} < \frac{7}{5} < \frac{41}{29} < \sqrt{2} < \frac{99}{70} < \frac{17}{12} < \frac{3}{2}$$

I know we haven't proved anything yet about the fractions in the sequence that follow the first six:

$$\frac{1}{1}, \quad \frac{3}{2}, \quad \frac{7}{5}, \quad \frac{17}{12}, \quad \frac{41}{29}, \quad \frac{99}{70}$$

but it looks, then, as if the very first fraction is the smallest of all the fractions in the sequence, while the second of the fractions is the largest of them all.

Another interesting observation.

If this the case, all the fractions, except for $\frac{1}{1}$, are greater than or equal to $1.4 = \frac{7}{5}$ and less than or equal to $1.5 = \frac{3}{2}$.

It would appear that way.

So the fractions alternate between being underestimates and overestimates of $\sqrt{2}$ simply because of the plus and minus property.

Yes. Actually, it is very handy to have the fractions alternate between being underestimates and overestimates of $\sqrt{2}$ because we can use them to place $\sqrt{2}$ into narrower and narrower intervals of the number line.

As if you were squeezing $\sqrt{2}$.

You could say that. For example, taking only the fraction $\frac{7}{5}$ on the left of $\sqrt{2}$ and the fraction $\frac{3}{2}$ to the right of $\sqrt{2}$ we get the inequality

$$\frac{7}{5} < \sqrt{2} < \frac{3}{2}$$

which you may recognize.

Something I mentioned earlier?

Yes, you said that 1.4 squared is less than 2 but that 1.5 squared is greater than 2.

A pure accident.

Maybe, or a sign of deep mathematical intuition.

Without doubt! So now we can improve on this and say that

$$\frac{41}{29} < \sqrt{2} < \frac{99}{70}$$

Correct. We cannot say, at least not yet, exactly how close $\frac{99}{70}$ is to $\sqrt{2}$ in terms of fractions or in decimal terms because we don't know how to calculate

$$\frac{99}{70} - \sqrt{2}$$

But we could, if we could only find the fraction that is the same as $\sqrt{2}$.

Certainly, but this fraction is eluding us at the moment. Still we can *estimate* how close the fraction $\frac{99}{70}$ is to $\sqrt{2}$.

How?

Let us look at the interval between $\frac{41}{29}$ and $\frac{99}{70}$ under the microscope, as it were.

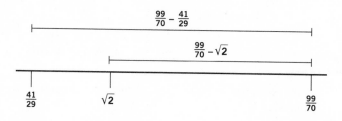

It may not strike you as a remarkable observation, but we can now at least say that the distance between $\sqrt{2}$ and $\frac{99}{70}$ is less than the length of the interval from $\frac{41}{29}$ to $\frac{99}{70}$, in which $\sqrt{2}$ resides.

This seems obvious from the picture you have just drawn.

In fact, since we know that $\frac{99}{70}$ is greater than $\sqrt{2}$ but closer to it than the fraction $\frac{41}{29}$, we may say that $\frac{99}{70}$ is within *half* the length of the interval between $\frac{41}{29}$ and $\frac{99}{70}$.

Of course; simple but clever.

The length of the interval between $\frac{99}{70}$ and $\frac{41}{29}$ is

$$\frac{99}{70} - \frac{41}{29} = \frac{(99 \times 29) - (70 \times 41)}{70 \times 29} = \frac{1}{2030}$$

Pretty narrow.

Since 2030 is bigger than 2000, we can say that the fraction $\frac{1}{2030}$ is smaller than $\frac{1}{2000}$. So the length of the interval is less than

$$\frac{1}{2000} = 0.0005$$

Less than 5 ten-thousandths of a unit.

Yes. This means that

$$\left(\frac{99}{70} - \sqrt{2}\right) < \frac{0.0005}{2} = 0.00025$$

an estimate that shows with minimum computation that $\frac{99}{70}$ is within 0.00025 of $\sqrt{2}$.

Very smart.

Why don't you use your rule to show that the next two terms in the sequence

$$\frac{1}{1}, \frac{3}{2}, \frac{7}{5}, \frac{17}{12}, \frac{41}{29}, \frac{99}{70}, \ldots$$

are the fractions

$$\frac{239}{169} \quad \text{and} \quad \frac{577}{408}$$

respectively, and verify that $239^2 - 2(169)^2 = -1$, with $577^2 - 2(408)^2 = 1$?

So the next two fractions also follow the plus or minus 1 pattern.

Yes, but these two facts prove nothing about the remaining fractions.

I realise this.

However, you might like to use these two new arrivals to show that

$$\frac{1}{1} < \frac{7}{5} < \frac{41}{29} < \frac{239}{169} < \sqrt{2} < \frac{577}{408} < \frac{99}{70} < \frac{17}{12} < \frac{3}{2}$$

A further homing in on where $\sqrt{2}$ lives. It's very impressive how much can be said with just simple mathematics.

True, but it does help to have good observations to work on.

A lesson I've learned from all of this in relation to the search for a fraction exactly matching $\sqrt{2}$ is that it could be an awfully long search.

Why?

Well, we have just shown that the leading six fractions of the sequence

$$\frac{1}{1}, \frac{3}{2}, \frac{7}{5}, \frac{17}{12}, \frac{41}{29}, \frac{99}{70}, \frac{239}{169}, \frac{577}{408}, \ldots$$

provide successively improving approximations of $\sqrt{2}$, and I suspect that the fractions further out this sequence do even better.

For the sake of argument we'll grant for the moment that they do.

Judging from the numerators and denominators of the first eight terms, I'm guessing that the numerators and denominators grow longer and longer as we move further out the sequence.

Another interesting observation that we might discuss in more detail later. But for the moment, where is this line of reasoning taking you?

Well, it suggests that the actual fraction exactly matching $\sqrt{2}$ may also have a very large numerator and denominator and, if so, searching for it could take a very long time.

You have a point.

For example, even if $\sqrt{2}$ were the fraction $\frac{99}{70}$ with its very modest numerator and denominator, I would have to search as far as the ninety-ninth perfect square before hearing a click.

And if $\sqrt{2}$ were the fraction

$$\frac{3515043237929985687829131076921717644468626388411}{2485510909704211897294694733728148710290930026229}$$

... which it isn't, by the way, although it is very very close, you could be ...

... searching for the rest of my life.

What the Ancients Knew

So are you going to give up on the search idea?

> Maybe, but I'd still like to test just a few more squares in the hope of getting lucky, even though I now realize that it is a most impractical method.

And one that would not produce a positive result no matter how far you, or countless millions of others armed with all the computing power in the world, were prepared to search.

> What did you say?

That you would never succeed. Your search would be in vain.

> Are you telling me that, of the infinity of fractions lying between 1.4 and 1.5 there is not one that squares to give 2 exactly?

That's correct. There *isn't* a fraction between these two numbers that squares to give 2 exactly.

> But if there isn't such a fraction—and how on earth could you be convinced that there isn't—what kind of number is it that, when squared, gives 2? Or are you going to say that there is no such number?

Ah! A moment of truth has arrived. These crucial questions, which our opening geometrical demonstration has forced upon us, are ones we must attempt to answer.

> Am I to understand that $\sqrt{2}$ is definitely not a fraction?

Yes, there is no *rational* number that, when squared, gives 2. Integers and fractions are known collectively as rational numbers. Put another way, there is no rational number that measures the length of the diagonal of a unit square.

> Incredible! Of the infinity of fractions between 1.4 or $\frac{7}{5}$ and 1.5 or $\frac{3}{2}$ you are absolutely certain that there isn't a single one of them that squares to give 2 exactly?

Absolutely.

> But how do you know for certain that such a fraction doesn't exist?

I know because the Ancient Greeks proved that it is impossible. I will show you one beautiful numerical proof.

> It must be a very deep proof that shows that there isn't a number that squares to 2 exactly.

No, that would be going too far! I'm definitely not saying that there isn't a "number" whose square is exactly 2. All I am saying is that there isn't an integer or a fraction which when squared gives 2 exactly. There is a difference.

But what other numbers are there besides the rational numbers, as you have just called them?

This is the mystifying point about the length of the diagonal of a unit square. It presents us with a paradox—an apparent contradiction—about the nature of numbers.

So all along you have known that my search was futile.

Futile in its ambition but not otherwise. I didn't want to give the game away. You are not the first to believe with complete conviction that there must be a fraction, however hard it might be to find, that squares to give 2 exactly. Besides, I wanted you to enjoy exploring and discovering, to experience the pleasure of finding things out for yourself.

I must gather my thoughts. I would not deny that the diagonal of the unit square has a length. In fact, this length is obviously greater than 1 unit, and as we know, less than 1.5 units. Yet you tell me that the length of this diagonal cannot be expressed as a unit plus a certain fraction of a unit.

That's right. While the rational numbers are perfectly adequate for the world of commerce, they are not up to the task of measuring the *exact* length of a diagonal of a unit square. No matter how close a rational number may come to measuring the length, there will always be an error, microscopically small perhaps, but nevertheless an error. Always. The ancient way of putting this was to say that the diagonal and side of a square are *incommensurable*.

So if we were to insist on thinking that *all* numbers are the ones with which we are familiar, namely the rationals, then we'd be forced to say that there is no number of units that measure this diagonal, or that there is no number whose square is 2.

Yes, but why restrict ourselves to such a viewpoint?

It seems natural.

Maybe, but perhaps it seems this way simply because most people's experience is limited to dealing with rational numbers. However, as you have said, if we were to insist on maintaining that rational numbers are the only type of number, then we'd have to be prepared to live in a world where there are lengths which are not measurable and where certain numbers have no square roots.

So we must accept that there are other types of numbers.

For mathematicians, the proof that no unit plus a fraction of a unit can hope to exactly measure the diagonal forces us to broaden our notions of what constitutes a number. When we do this, the paradox surrounding $\sqrt{2}$ simply dissolves.

So what "number" measures the diagonal of a unit square?

The one whose square is 2 and that we denote by $\sqrt{2}$. We admit the existence of this number because it makes its presence necessary by being the length of a legitimate quantity—the diagonal of a unit-square.

> So the length of any side of the internal square we talked about at the beginning is simply $\sqrt{2}$, with no need for further elaboration.

Yes. $\sqrt{2}$ is a number between 1.4 and 1.5 that is not a rational number but that, when squared, gives 2. As we have already said, $\sqrt{2}$ is defined by the equation $\sqrt{2} \times \sqrt{2} = 2$, which is the mathematical way of saying that $\sqrt{2}$ is the positive number that squares to give 2.

> So $\sqrt{2}$ is a new type of number.

Yes, new or different, but we have not proved this yet. Because it is not a rational number, it is called an *irrational* number. Not that there is anything unreasonable about it. It is so named because it cannot be expressed as the *ratio* of two integers in the way that the fractions are.

> So the word rational in "rational number" is used because of the word ratio, while the term "irrational" in connection with $\sqrt{2}$ is used because it cannot be so expressed.

Quite so. The number $\sqrt{2}$ is as real as any fraction. In fact, $\sqrt{2}$ is just one of an infinite number of irrational real numbers that exist "outside" the realm of the rational numbers.

> Can you show me some other irrational numbers?

Yes, the positive square roots of each of the other numbers missing from the list of perfect squares we made out some time ago can also be shown to be irrational numbers.

> This means that
>
> $$\sqrt{2}, \quad \sqrt{3}, \quad \sqrt{5}, \quad \sqrt{7}, \quad \sqrt{8}, \quad \sqrt{10}, \ldots$$
>
> are all irrational numbers.

Yes.

> This is why there is an infinity of these numbers.

Certainly, but the collection of irrational numbers contains not just all these *surds*, as they are sometimes called, but a whole galaxy of other weird and wonderful numbers, the most famous being π.

> Ah, π! The ratio of the length of the circumference of any circle to the length of its diagonal. I thought that π was the fraction $\frac{22}{7}$.

This is only an approximation of its true value, just like $\frac{7}{5}$ is an approximation of $\sqrt{2}$.

> Reality is a lot more complicated than I naïvely thought.

Perhaps we should say that the world of mathematics is a lot more complicated than one might think at first. However, speaking of reality, the collection of rational numbers and the collection of irrational numbers between them constitute the set of *real* numbers.

> The idea that there are new specimens of numbers other than the "usual ones" used in arithmetic takes a little getting used to.

You're not the first person who was more than a little perplexed by these new numbers. The minds of the Ancient Greeks were bewildered when these irrational numbers thrust their existence upon the Greeks' consciousness. Legend has it that they were positively perturbed by the intrusion of these new quantities into their reality. They experienced an intellectual and philosophical crisis.

> They did? Why?

Well, there was a brotherhood of Pythagoreans, followers of the famous philosopher and mathematician Pythagoras, which was devoted to the pursuit of higher learning, in particular mathematics. They were very well respected and considered to know all that there was to know. They believed that everything could be quantified by the familiar rational numbers.

> A reasonable enough belief, or should I say a rational belief?

Yes, a very justifiable one. After all, these numbers are the only ones needed for commercial transactions, and they are equally adequate at describing various other physical phenomena. They also suffice for most measuring purposes that occur in practice.

> Even though they cannot be used to give the measure of the diagonal of a unit square.

Yes, the issue about the new nature of $\sqrt{2}$ and its cousins, $\sqrt{3}$, $\sqrt{5}, \ldots$ was a theoretical one rather than a concern with "practical" measurement. The Greeks were fully aware that even if fractions could not measure the diagonal of a unit square exactly, they could measure it to any desired degree of accuracy. For example, a length of

$$\frac{577}{408} = 1 + \frac{169}{408}$$

units measures the "true" length of the diagonal to well within a hundred-thousandth of a unit.

> Which is less than one-hundredth of a millimeter if the unit is a meter.

[See chapter note 3.]

Tablet 7289 Yale
Collection

There is evidence that this approximation to $\sqrt{2}$ was known to the Babylonians around 1600 B.C. This is many centuries before the Ancient Greeks whom I mentioned, because a Babylonian tablet from that time gives 1; 24, 51, 10 as an approximation to $\sqrt{2}$.

What does 1; 24, 51, 10 stand for?

It's shorthand for

$$1 + \frac{24}{60} + \frac{51}{60^2} + \frac{10}{60^3}$$

The Babylonians used a sexagesimal system.

What is this when it is simplified?

The fraction

$$\frac{30547}{21600}$$

which, as you can see, is not $\frac{577}{408}$ exactly.

Why, then, is it thought that they knew of "our" $\frac{577}{408}$?

Because in base 60,

$$\frac{577}{408} = 1; 24, 51, 10, 35 \ldots$$

it is suspected that they just truncated (shortened) the sexagesimal expansion of this fraction.

To three places, as we'd say.

Yes.

How did the Babylonians find such approximations?

It is not exactly known, but there is speculation that they knew of a method of approximation.

Was it a different method from the one using the sequence of fractions we have discovered?

It is related to this but faster.

Faster sounds interesting.

Accelerated, we might even say. This method also gave them 1; 25 as an approximation of $\sqrt{2}$. Convert 1; 25 to base 10 to see what it is.

I'll try. Since they used a sexagesimal system

What is 1; 24 as a fraction?

$$1; 25 = 1 + \frac{25}{60} = \frac{85}{60} = \frac{17}{12}$$

this fraction, just like $\frac{577}{408}$, is in our sequence.

It is indeed, the fourth in the sequence. It is not as good an approximation of $\sqrt{2}$ as $\frac{577}{408}$, which is the eighth entry in the same sequence. As we said before, it doesn't do a bad job of approximating $\sqrt{2}$.

So these Mesopotamians must have known their mathematics.

And quite a bit more, by all accounts.

I should be able to verify for myself that $\frac{577}{408}$ approximates $\sqrt{2}$ as closely as you say.

You should and, what is significant, without knowing anything about the decimal expansion of $\sqrt{2}$.

Hmm. I didn't appreciate this point before.

I didn't emphasize it prior to this.

Please remind me of how exactly I would begin to go about this verification.

Recall that the fraction $\frac{239}{169}$ is the one before $\frac{577}{408}$ in our short sequence and that it underestimates $\sqrt{2}$, whereas $\frac{577}{408}$ overestimates it.

I think I remember now. Since $\frac{577}{408}$ is closer to $\sqrt{2}$ than $\frac{239}{169}$ is, its distance from $\sqrt{2}$ is less than half the distance between these two fractions.

Yes. This distance is $\frac{1}{68952}$, as you can check.

And what do we say now?

Since $50000 < 68592$, we say that $\frac{1}{68592}$ is less than $\frac{1}{50000}$. Hence the length of the interval is less than one fifty-thousandth of a unit, and so $\frac{577}{408}$ is within a hundred-thousandth of a unit of $\sqrt{2}$.

So we're done.

Yes. Maybe now is a good time to use what we know to get some idea of the leading digits in the decimal expansion of $\sqrt{2}$.

How are you going to do this?

Convert the fractions in the inequality

$$\frac{239}{169} < \sqrt{2} < \frac{577}{408}$$

to decimal form.

Using a calculator I hope.

Yes, because in theory this is something we can do ourselves by hand.

And so we are free to use a calculator to save time.

We get

$$1.4142011834319526627\ldots < \sqrt{2} < 1.4142156862745098039\ldots$$

working to twenty decimal places.

Some calculator! I see that both expansions agree to four decimal places. Is it safe so to say that

$$\sqrt{2} = 1.4142\ldots$$

which, I think, is pretty good?

It is. And because we did everything ourselves I think we can take a bow. We'll have fun improving on all of this at a later stage.

But to get back to the Ancient Greeks. You were saying that it is the *nature* of numbers that was of primary interest to these learned men.

Indeed. Such was their conviction that the rational numbers described all of nature exactly that their motto was, "All is number."

By which they meant the rational numbers.

Yes.

I'm glad to see I'm in good company.

You could certainly say that.

So they had their colors well and truly nailed to the mast.

This proclamation took on the status of an incontrovertible truth. It became a creed.

Oh! The discovery of the existence of $\sqrt{2}$ must have come as a shock.

A most unwelcome one we are told, because it challenged their cherished belief about the nature of numbers.

They took this whole business about numbers really seriously then?

I don't know how true much of the early lore surrounding the discovery of $\sqrt{2}$ is, but one story has it one of the brotherhood leaked the news that all was not well with the accepted dogma. For this breach of faith he was taken on a sea trip and cast overboard.

[See chapter note 4.]

You're kidding me!

Well, if it is true, it goes toward answering your question as to how seriously they took their mathematics.

So his number was up!

As fate would have it, the number $\sqrt{2}$ is referred to in some quarters as Pythagoras's constant.

I wonder what the Pythagorean brotherhood would have to say about that. But how did they come to know for certain that $\sqrt{2}$

is not a rational number? Surely they must have thought that $\sqrt{2}$ is actually a rational number but that they simply lacked the means to find it?

Perhaps they came upon numerical evidence similar to what you found in your search, but I don't know. I do know that their main mathematical focus was geometry.

Of course, the famous theorem of Pythagoras.

Actually, it may have been this very theorem that first brought $\sqrt{2}$ to the attention of the Ancient Greeks.

So they were the cause of the downfall of their own philosophy that "all is number."

You could say that. Coming back to what we were saying about searching, these clever Greeks would have known that the search method is one that, no matter how many perfect squares may have been checked, still leaves an infinite number of possibilities to be tested.

I hope they figured that out faster than I did!

I'm sure they were fully aware that any finite quantity, no matter what its size, is as nothing against the backdrop of infinity.

Still, they must have suspected right from the very moment the $\sqrt{2}$ problem reared its head that their doctrine of number was in trouble.

I'd be inclined to agree: they must have known that the doctrine wasn't as all-embracing as they originally proclaimed. It may be that some were really intrigued that $\sqrt{2}$ does not dwell in the infinite realm of rational numbers but is something that is "outside of it," as it were. Certainly minds over the centuries have been charmed by this aspect of numbers.

Did the Ancient Greeks find the proof you mentioned fairly soon after observing that there was more to $\sqrt{2}$ than meets the eye?

As far as I know, quite a stretch of time elapsed, about 300 years or so, before someone found an argument that turned suspicion into fact and established the irrationality of $\sqrt{2}$ once and for all. However, I don't know if the argument described by Euclid, which I am about to show you, was the first because there are many ingenious proofs of the irrationality of $\sqrt{2}$.

Elements X, §115a

But they must all be very difficult. It cannot be easy to be convinced that of the infinity of rational numbers, not one squares to give 2.

Not at all. The proof we are about to discuss is a magnificent *reductio ad absurdum* argument.

Which means?

In this case, you assume that there is a rational number that, when squared, gives 2, and then you show that this assumption leads to a contradiction or, put another way, reduces to something absurd. This form of logic—bequeathed to us by these Greek scholars—has been used ever since throughout mathematics.

If you arrive at a contradiction, you say that the assumption you made at the start is the cause of the trouble.

Yes, and you conclude that it must be false and must be abandoned.

So if the assumption is false then its opposite is true?

Precisely.

After thinking that $\sqrt{2}$ *must* be a fraction and having been frustrated in a vain search, I cannot wait to see this proof of irrationality. It must be a wonderful mathematical work of art.

A work of art indeed. Bertrand Russell once said, "Mathematics, rightly viewed, possesses not only truth but supreme beauty . . . sublimely pure, capable of a stern perfection as only the greatest art can show." Judge for yourself whether the proof merits this accolade.

(1872–1970)

CHAPTER 2

Irrationality and Its Consequences

To prove that $\sqrt{2}$ is irrational, assume the opposite.

You mean assume that it is a rational number?

Yes.

If $\sqrt{2}$ is rational, then, as we said before, there are two natural numbers, say m and n, such that

$$\sqrt{2} = \frac{m}{n}$$

Right?

Correct.

And when we square both sides of this equation, we get

$$2 = \frac{m^2}{n^2}$$

How am I doing?

Very well.

I'm afraid that when I multiply this equation across by n^2, my contribution toward this immortal proof will be at an end.

Perhaps, but it is exactly the way to start the proof.

Thank you. It was fun while it lasted!

But before we multiply across by n^2, I want to say a little about m and n. If m and n have factors in common, then each of these factors can be removed from both at the outset.

How?

Simply by canceling above and below the line as one would the common factors 3 and 4 in $\frac{36}{48}$, to get

$$\frac{36}{84} = \frac{3 \times \cancel{3} \times \cancel{4}}{\cancel{3} \times \cancel{4} \times 7} = \frac{3}{7}$$

Do you agree?

> I see; it seems like nothing more than common sense.

So from the outset we may assume, without any loss of gener-
ality, as they say, that the natural numbers m and n have no
factors in common other than the trivial factor 1.

> No factors other than 1 in common. I suppose every pair of
> natural numbers has the factor 1 in common.

Yes, because the number 1 is a factor of every number, it is
called a *trivial* factor.

> You are saying, then, that we may assume at the beginning that
> the numerator m and the denominator n have no nontrivial
> factors in common.

Exactly.

> This seems reasonable.

In this case, the fraction $\frac{m}{n}$ is said to be in *lowest* or *reduced*
form.

> So the fraction $\frac{3}{7}$ on the right-hand side of the example is in
> lowest form because its numerator and denominator have no
> factors in common.

Precisely. By cancelling the common 3 and 4, the fraction $\frac{36}{48}$
reduces to the fraction $\frac{3}{7}$. Since $\frac{3}{7}$ cannot be reduced any further,
it is in lowest form.

> Well and truly, I would say.

Now that we have this detail out of the way, let us multiply

$$2 = \frac{m^2}{n^2}$$

across by n^2, as you were about to do, to get, after interchanging the terms on each side,

$$m^2 = 2n^2$$

This is an equation with which you are already quite familiar.

> How can I forget, back when I foolishly expected to find small
> values for m and n, which would make it true.

There was nothing foolish in what you did. Look on the bright
side. You made a discovery that led us to better and better
approximations of $\sqrt{2}$.

> And because of what you are about to prove, approximations
> of $\sqrt{2}$ are "as good as it gets" using fractions.

Too true. Besides, because of all your previous hard work, you'll have a much greater appreciation of this proof, which is now entering its serious phase.

> I must pay full attention then. I don't want to miss a detail, particularly because I want to understand the thought process of the ancient who first proved this remarkable result so long ago.

The equation $m^2 = 2n^2$ tells us that m^2 is an even number.

> Let me think why. Oh, I see why; because it is twice n^2 and so must be even, since twice any number is even.

Correct. Now that you know m^2 is an even number, can you say whether m is even or odd?

> Could it be either?

No. Since the square of an odd number is always odd, it must be that the natural number m is also an even number.

> So only even numbers square to give even numbers. Simple when you are told the reason.

You might like to convince yourself sometime, why the square of an odd number is also an odd number. Anyway, if m is even, it means that it is twice some other natural number, p, say.

$$(2k - 1)^2 =$$
$$4(k^2 - k) + 1$$

> An example, please.

For example, $14 = 2 \times 7$. Here m is the 14 and p is the 7.

> Got it.

So, $m = 2p$ for some natural number p.

> Right.

Now, substituting $2p$ for m in $m^2 = 2n^2$ gives $(2p)^2 = 2n^2$ or $4p^2 = 2n^2$.

> With you so far.

Canceling the 2 common to both sides of the very last equation gives, after interchanging the terms on each side,

$$n^2 = 2p^2$$

which you may notice is exactly like the original equation $m^2 = 2n^2$.

> With n where m was and p where n was?

Exactly. Can you say what this latest equation tells us about n?

> More thinking. Doesn't it tell us that n^2 is also an even number, just like m^2?

It does. What is an implication of this?

> I suppose for the same reason as before it means that n is also an even natural number.

It does indeed. The proof is finished.

> What, so soon? I must be missing something. Tell me why the proof is finished.

Because we have arrived at a contradiction.

> I need time to see where it is.

Take as much time as you need.

> I see it, now. We now know that m and n must both be even numbers.

Yes, and what is wrong with this?

> Didn't we say at the start that the numbers m and n have no factor in common?

Other than 1, we did.

> So this is the contradiction then?

It is. We began by assuming that

$$\sqrt{2} = \frac{m}{n}$$

with the natural numbers m and n having no common factor other than 1. But the argument we have just given shows convincingly that this assumption forces m and n to be both even.

> Which means that m and n have 2 as a common factor. But this contradicts the original assumption that they have no factors in common except 1, which is a pain to have to keep mentioning.

Now you have it; the assumption "reduces to an absurdity."

> So the original equation can never be true for *any* natural numbers m and n.

Never! Hence we may declare with certainty that $\sqrt{2}$ is an irrational number.

> Magnificent! And it is not hard when somebody shows you how it is done.

That is part of its charm.

> And so short. I was expecting a long argument.

It's a real gem. A classic example of the *reductio ad absurdum* argument.

> That was fascinating! I almost feel like a mathematician myself now.

That's wonderful, I'm glad you appreciate it. Let's summarize the proof in a bare-bones fashion just to show how elegant it is.

> Like you might see in a mathematics text.

Yes. But you'll be able to see the sense in it now. By the way, the symbol $\Rightarrow$ just means "implies that." So here is a streamlined version of the proof: assume that there are natural numbers m and n with no nontrivial factor in common, such that $\sqrt{2} = \frac{m}{n}$. Then

$$\sqrt{2} = \frac{m}{n} \Rightarrow 2 = \frac{m^2}{n^2}$$

$$\Rightarrow m^2 = 2n^2$$

$\Rightarrow m^2$ is an even natural number.

$\Rightarrow m$ itself is an even natural number.

$\Rightarrow m = 2p$ where p is another natural number.

$$\Rightarrow (2p)^2 = 2n^2$$

$$\Rightarrow n^2 = 2p^2$$

$\Rightarrow n^2$ is an even natural number.

$\Rightarrow n$ itself is an even natural number.

$\Rightarrow m$ and n are both even natural numbers.

$\Rightarrow 2$ is a common factor of m and n.

$\Rightarrow$ a contradiction since m and n have no common factor other than 1.

Really concise; it makes ordinary language seem very long-winded by comparison. I can see what Russell meant by "stern perfection."

A mere dozen lines. In terms of calculation, the mathematical content of this proof is completely elementary; it is the reasoning used that is spectacular.

Some of those Greeks were really smart!

We might be justified in saying that the "dawning" on someone that it is impossible to express $\sqrt{2}$ as a rational number was a thought of pure genius. Equally, the first one to prove this impossibility was also a genius. This proof so enthralled the mathematician G. H. Hardy as a boy that, from the moment he read it, he decided to devote his life to mathematics.

And did he?

He certainly did. He became one of England's great mathematicians of the twentieth century. When he believed that his mathematical creativity was at the end, he wrote a delightful book called *A Mathematician's Apology*.

Apology?

In the sense of an *apologia*. It was his view that the job of a genuine mathematician is to do mathematics rather than talk

G. H. Hardy
(1877–1947)

about it. It saddened him at the time of writing that he could no longer do the former to the high standard he demanded of himself.

So he was very hard on himself?

To judge from what he wrote. He also had a very pure view of mathematics. An oft quoted statement of his is, "Beauty is the first test. There is no permanent place in the world for ugly mathematics."

Sounds nice, but is this true?

Whether it is or not, he seemed proud of the fact that none of the mathematics he created ever found an application.

This seems to me like a strange thing to say.

One that raised hackles. It provoked one eminent scientist, a fellow Englishman, to exclaim, "From such cloistral clowning the world sickens."

Ouch!

Consequences of the Irrationality of $\sqrt{2}$

Earlier, you said the ancients used to say that the side and diagonal of a square are *incommensurable*. Would you spell out what they meant by this?

They meant that the side of a square and its diagonal cannot both be measured exactly with the same ruler, no matter how fine its markings.

Hard to imagine.

I know, but if you have a ruler that measures the side exactly (meaning that the two endpoints of the side coincide with two markings of the ruler), then that same ruler, when placed along the diagonal so that one of its markings coincides with the initial point of the diagonal, will have the endpoint of the diagonal lying between some two adjacent markings.

> I don't really believe it. Always *between* and never coinciding?

Inescapable! And if the ruler measures the diagonal exactly, it will fail to measure the side exactly. If you had at your disposal a new ruler with ultra-close markings, or even an infinity of such rulers with any conceivable separation between successive markings, it would make no difference.

> Is it difficult to explain why?

It is very easy to see using the tiniest amount of algebra.

> Show me, please. I enjoyed the last argument even though it used letters.

The way to show that it is impossible is to assume that it is possible and arrive at a contradiction.

> Just like the proof of the irrationality of $\sqrt{2}$.

Just so; and as we'll see, it is this fact that decides emphatically that the side of a square and its diagonal cannot both be measured exactly with the same ruler. So we'll assume that there is some ruler that can exactly measure both the side and the diagonal.

> It's hard to believe that such a ruler doesn't exist.

To begin, we can let the length of the side of the square be 1 unit. Then the length of the diagonal, as we know, is $\sqrt{2}$ units. Let u stand for the distance between any two successive marks on the ruler. Since this ruler is supposed to be able to measure both lengths exactly, the length of the side is made up of a whole number of u's. Call this natural number n. This means that

$$1 = nu$$

Similarly, the length $\sqrt{2}$ of the diagonal is also made up of a whole number of u's. Call this natural number m.

> So

$$\sqrt{2} = mu$$

> in this case.

Definitely.

> I'm with you so far. But what do you do now?

Something simple but clever. We divide the second equation by the first to get

$$\frac{\sqrt{2}}{1} = \frac{mu}{nu}$$

Can you see what to do now?

Cancel the u's above and below the line?

Yes. That this can be done is very significant. It shows that the size of u, separating successive markings, does not influence what follows. What do you get?

That

$$\sqrt{2} = \frac{m}{n}$$

So what?

Look carefully at what this simple equation is saying. What would the ancient Greeks say?

This equation is saying that $\sqrt{2}$ can be expressed as a fraction. But this is *false*!

And so?

Since the assumption that both the side and the diagonal of a square can be exactly measured with the ruler leads to a contradiction, it is also false.

So we are forced to the conclusion, however strange it may seem, that these two sides are incommensurable.

I can't argue with that.

Variation on a Theme

Can you tell me some other concrete consequences of the irrationality of $\sqrt{2}$?

I can. Imagine a rectangular room having dimensions where the ratio of the length of the longer side to the shorter side is exactly $\sqrt{2} : 1$. If we take the length of the shorter side as our unit, then the rectangle looks something like this:

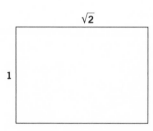

Then this rectangle cannot be tiled with square tiles, no matter how small the tiles.

> I presume there is no question of grouting between the tiles.

Joking aside, absolutely not!

> And not even the hint of a leftover gap along either side?

There must be no cutting of the small tiles to fill any gap, no matter how minuscule.

> I suppose the explanation of why this tiling is unachievable is no harder than the one just given?

The proof that any such tiling is impossible follows exactly the same lines as the argument just described. You might like to think about it. We can come back to it later if you like.

> Will do. Where might you come across such a rectangle?

Rectangles with this exact shape have a unique and very special property.

> They do?

Yes. The ISO A4 sheet (a metric standard paper size) is a rectangle of this type. Ideally, it's supposed to have the length of its longer to its shorter side in the *exact* ratio $\sqrt{2}:1$.

> I have an A4 pad of paper here.

Well, look at the cover and see what is said about dimensions.

> It says that the pad measures 297 millimeters by 210 millimeters.

At first sight, rather odd dimensions, wouldn't you say?

> Now that you mention it. What's the reason?

Work out the ratio of the longer side to the shorter side.

> I get

$$\frac{297}{210} = \frac{99}{70}$$

So?

Ah, how soon we forget!

> Oops! Stupid me. This fraction is an old friend. It's the sixth in the sequence
>
> $$\frac{1}{1}, \ \frac{3}{2}, \ \frac{7}{5}, \ \frac{17}{12}, \ \frac{41}{29}, \ \frac{99}{70}, \ \frac{239}{169}, \ \frac{577}{408}, \ \ldots$$

of approximations to $\sqrt{2}$ that we hit upon some time ago.

None other. You may recall we showed that the fraction $\frac{297}{210} = \frac{99}{70}$ is well within 0.00025 of $\sqrt{2}$.

Therefore, $\frac{297}{210}$ is a good approximation of $\sqrt{2}$.

Less than one four-thousandth of a unit in terms of meters is less than one-quarter of a millimeter.

So we are witnessing a connection between a term in this sequence and the paper industry?

Yes.

You are saying that these dimensions are purposely chosen so that the ratio of the longer side of an A4 sheet to its shorter side is very nearly in the ratio $\sqrt{2}:1$.

As I said, ideally, they should be in this exact ratio but, as we now know, this cannot be achieved with each side being an exact natural number of millimeters in length.

No doubt there are good reasons for insisting on trying to achieve this ratio.

The reason is a practical one. In this case, and in this case only, the sheet can be folded along its longer side to give two smaller rectangles, for each of which the ratio of the longer to the shorter side is *again* $\sqrt{2}:1$.

It must be this way and no other way. Really intriguing!

And because this ratio cannot be achieved with both sides being integer multiples of a basic unit, you wouldn't know whether to describe nature as subtle or just plain contrary.

But she's certainly interesting. Who spotted this first, and is it really useful?

[See chapter note 1]

Well, I've read that it makes paper stocking and document reproduction cheaper and more efficient.

So even $\sqrt{2}$ has a "commercial side" to it.

Georg Christoph Lichtenberg, University of Göttingen, Germany, 1742–1799

According to the same source, "the practical and aesthetic advantages of $\sqrt{2}$ aspect ratio for paper sizes were probably first noted by the physics professor Georg Christoph Lichtenberg in a letter he wrote on October 25, 1786, to one Johann Beckmann." As a result of his observation, the ratio $\sqrt{2}:1$ is called Lichtenberg's ratio in some quarters. Now there's a little bit of history for you.

And much more recent than I would have expected. Is it hard to show why $\sqrt{2}:1$ is the only ratio that works?

No. I'll show you why in a few moments. But to continue: each of the new sheets thus formed has the same property as the parent.

And this is important?

Yes, because it means that each of the smaller sheets can in turn be folded in two to give more sheets of the same type but smaller, and so on if required.

Very nice!

Two A4 sheets are the offspring of a single A3 sheet, and in turn, two A3s come from one A2. The first in the line of the A-series is an A0 sheet whose dimensions in millimeters are 1189 × 841. This is approximately a square meter in area.

Let me check this:

$$841 \times 1189 = 999{,}949$$

which is almost the 1,000,000 millimeters squared contained in a square meter.

Yes. The area of an A0 sheet is 51 millimeters squared short of the full square metre. This error might appear significant, but as a percentage of 1,000,000, it is a mere 0.0051%.

A tiny percentage.

Why don't you fold an A4 page in two and check that the ratio of the longer to the shorter side is approximately $\sqrt{2}:1$. Easier still, place two A4 sheets side by side lengthways and check that the ratio of the longer to the shorter side is approximately $\sqrt{2}:1$.

I suppose I don't even have to do it physically, because if I place a long side of each A4 side by side like this:

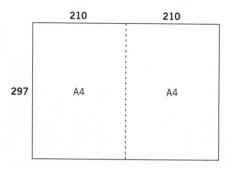

I must end up with a rectangle with long side measuring 2 × 210 = 420 millimeters and short side measuring 297 millimeters.

Just so. Why don't you check how close the fraction $\frac{420}{297}$ squared is to 2.

Ah, you're testing to see if I can still remember how to do this.

Never! You might as well reduce the fraction to its lowest form of $\frac{140}{99}$ before you start.

Oh right. The calculation

$$\left(\frac{140}{99}\right)^2 = \frac{19600}{9801}$$

$$\stackrel{!}{=} \frac{19602 - 2}{9801}$$

$$= 2 - \frac{2}{9801}$$

shows that the square of $\frac{140}{99}$ is within $\frac{2}{9801}$ of 2.

Well done! I see you haven't forgotten the idea of that trick we used so well earlier.

It took me a while to recall that I should write 19600 = 19602 − 2 so that I could divide 9801 into 19602 to get 2 exactly.

You'll be quite the old hand in no time!

You'll give me a big head. Anyway, we can say that the ratio of the long side of an A3 page to its shorter side is also approximately in the ratio $\sqrt{2}:1$.

If the A4 pages had the ideal ratio of $\sqrt{2}:1$, then combining two of them as above gives a sheet like this:

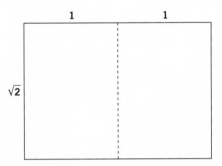

Here the long side measures 2 × 1 = 2 and the short side measures $\sqrt{2}$.

So the ratio of the long side of this big sheet to its short side is $2:\sqrt{2}$.

And is this the same as the ratio $\sqrt{2}:1$?

Ah, you want me to show that it is! You will have to give me a hint.

All right. Just remember that $\sqrt{2} \times \sqrt{2} = 2$.

Let me think a little while.

Take your time; inspiration will strike.

Let's hope so. There cannot be all that much I can do with such a meager expression as the ratio $2 : \sqrt{2}$.

As you say.

I think I see one thing I could try. I'll start simply by writing the 2 as $\sqrt{2} \times \sqrt{2}$.

Very good, because this step is the reverse of anything we have done before.

It is?

Up to now we have always replaced $\sqrt{2} \times \sqrt{2}$ by the simpler 2, but this time you are choosing to write the compact number 2 in expanded form, as $\sqrt{2} \times \sqrt{2}$.

I see what you mean. Doing this allows me to write

$$\frac{2}{\sqrt{2}} \overset{!}{=} \frac{\sqrt{2} \times \sqrt{2}}{\sqrt{2}}$$

Canceling a $\sqrt{2}$ above and below the line on the right I get this:

$$\frac{2}{\sqrt{2}} = \sqrt{2}$$

But what am I to do now?

A little bit of trickery. Since $\sqrt{2}$ is the same as $\frac{\sqrt{2}}{1}$, and since this is equivalent to the ratio $\sqrt{2} : 1$, you may say that this equation shows that

$$2 : \sqrt{2} = \sqrt{2} : 1$$

So, we're done.

Why?

Because, haven't we just shown, for the sheet of paper formed by gluing together two ideal A4 sheets, that the ratio of the long side to its short side is the ideal $\sqrt{2} : 1$?

You have, and so we get an ideal A3 sheet.

Fascinating!

Well done.

So now I know that if the ratio of the longer side of a sheet of paper to the shorter side is exactly $\sqrt{2} : 1$, then that piece of paper can be folded in half along its long side to give two smaller sheets of paper, to which the same thing can be done.

Correct. Each of these sheets has the same property: namely that the ratio of the longer to the shorter side is again $\sqrt{2}:1$.

> And so they too can, as you said earlier, be folded along their long sides to produce two smaller sheets that will also have the same property.

That's right.

> You also said earlier that rectangles of these dimensions are the *only* ones having this property.

I did. Only those rectangles, the length of whose long side to the length of the short side are in the ratio $\sqrt{2}:1$, have this property.

> I'd be interested to see why.

It will require some algebra.

> I thought it might!

Let us begin with a labeled diagram of a rectangle:

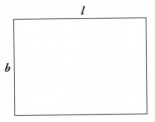

Here *l* stands for the length of the longer side and *b* stands for the length of the shorter side.

> I suppose one side has to be longer than the other for there to be any hope of success.

You are right to question my assumption that one side is longer than the other. Fortunately, it is fairly obvious that a square couldn't possibly work. The ratio of its sides are $1:1$ so that, when folded, the resulting two smaller pieces have sides in the ratio $2:1$, as their longer sides are twice as long as their shorter sides.

> Agreed. So one side of this special rectangle must be longer than the other.

It must, but that this is so emerges of its own accord from the demonstration. So let us divide this rectangle by folding perpendicular to its long side.

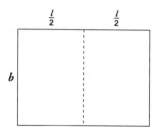

Now tell me the dimensions of the sides of each of the two smaller rectangles.

The shorter side has length $\frac{l}{2}$, while the longer side has length b.

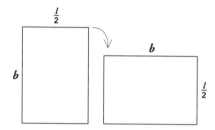

But how do you know the side of length $\frac{l}{2}$ is shorter than the other side of length b?

Well, it certainly looks shorter from the diagram. I suppose you'll want me to convince you that it must be.

I'm afraid so. If we had started out with a 12 × 4 rectangle so that $l = 12$ and $b = 4$, then $\frac{l}{2} = 6$ is still bigger than $b = 4$.

I can see that. But surely this is not the type of rectangle we're after? In your example, the ratio of the length of the longer side to the length of the shorter side is 12:4 or 3:1, while the corresponding ratio for either of the smaller rectangles is 6:4 or 3:2. A ratio of 3:2 isn't the same as a ratio of 3:1, so the ratios don't match.

You are right in what you say. No matter what l and b may be, we cannot have the length $\frac{l}{2}$ longer than b because otherwise we'd be trying to match the ratio $l:b$ to the ratio $\frac{l}{2}:b$, which is not possible.

Could $\frac{l}{2}$ be equal to b?

A good question, but no, for similar reasons. If $\frac{l}{2}$ were equal to b, then l would be equal to $2b$. This would mean that the rectangle had a $2:1$ ratio of its long side to its short side.

And the two small rectangles would actually be squares?

Yes, where the ratio of the long side to the short side is $1:1$.

Which rules out $\frac{l}{2} = b$ because the two ratios must match.

So b must be the length of the longer side in each of the smaller rectangles.

Well, I'm glad that is out of the way.

Notice that we have just figured out that, to get matching ratios, the length of the long side of the rectangle we start out with must be less than twice the length of the shorter side.

Having thought about it, it seems almost obvious.

Perhaps. Now to the juicy bit of the argument. For things to work we must have:

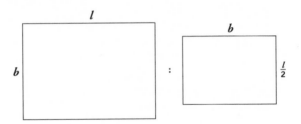

which means that we must have

$$l:b = b:\frac{l}{2}$$

or

$$\frac{l}{b} = \frac{b}{\frac{l}{2}}$$

Would you like to simplify this?

I'll try. In school we learned that to divide by a fraction, you turn it upside down and multiply.

Yes, and what do you get in this case?

We get that

$$\frac{l}{b} = 2\frac{b}{l}$$

But what do we do now?

Why don't we bring all the letters to one side by multiplying both sides of the equation by $\frac{1}{b}$. In that way you'll end up with the number 2 on its own on the right-hand side.

Doing this gives

$$\frac{l}{b}\frac{l}{b} = 2$$

if I remember my school algebra correctly.

Yes. This simple little equation says that

$$\left(\frac{l}{b}\right)^2 = 2$$

What do you make of this?

Hmm! Thinking time again. Is it saying that there is some fraction that when squared gives 2?

Definitely not, though it might look like it. Remember that, for us, the numerator and the denominator of a fraction are whole numbers. We never said that l and b had to be whole numbers.

Of course we didn't. In fact, doesn't this equation now tell us that l and b cannot both be whole numbers?

Absolutely correct, because if they were, then we'd have a fraction squaring to 2.

Which we know is impossible. Could they both be fractions?

No, because when you divide one fraction by another . . .

. . . you get another fraction. And a fraction squared can never give 2. So what is the equation telling us?

That

$$\frac{l}{b} = \sqrt{2}$$

or that

$$l:b = \sqrt{2}:1$$

which is what we set out to prove.

I'll have to gather my thoughts. Our mission was to discover a rectangle with the property that when it is folded along its longer side to give two smaller rectangles, the ratio of the longer to the shorter side of each of these smaller rectangles would be exactly the same as the ratio of the longer side of the original rectangle to its shorter side. Am I right?

Yes. And we succeeded in this. What have we discovered?

> That such rectangles exist and that the ratio of the length of the longer side of such a rectangle to the length of its shorter side *must* be exactly $\sqrt{2}:1$.

Yes. The above discussion doesn't say that *l*, or *b* for that matter, must be a specific number.

> It is only the ratio that matters and not the actual dimensions.

Exactly. The scale doesn't matter as long as the ratio is right.

> So things work for any rectangle whose dimensions are in the ratio $\sqrt{2}:1$, and they don't work otherwise.

That's the story from top to bottom.

But What About the Tiling Problem?

> Have you forgotten that we still must come up with an argument as to why it is impossible to tile the $\sqrt{2} \times 1$ rectangle with square tiles no matter how small?

Of course, I didn't actually prove that it is impossible. Why don't you try it?

> Me, prove it?

Yes. You'll have no problem, and it would round off this particular discussion rather nicely.

> So I must figure out why this rectangle cannot be tiled. Acting on previous experience, I'm going to assume that it can be tiled exactly, in the hope that I arrive at a contradiction.

A good plan. So are you about to start laying tiles?

> Mentally, but before I do, I had better decide on the dimensions of an individual tile. Since I want to allow for all possible square sizes, I'll let the tile have sides of length *s*. I'll draw one.

> Here *s* stands for one length and every length at the same time. Is this the right approach?

One for all. Very sophisticated.

> Let me suppose that I lay exactly *m* tiles along the long side of length $\sqrt{2}$, and exactly *n* tiles along the short side of length 1. The plan would look something like this:

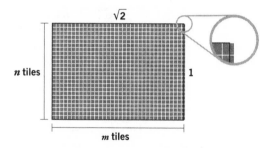

with no gaps between the square tiles.

I see.

Since there are no gaps, and each tile side has length s, it must be that

$$\sqrt{2} = ms$$

while

$$1 = ns$$

Is this not so?

Seems logical to me since m tiles, each of length s, measure exactly ms units, with n such tiles measuring ns units.

Now let me think. What did we do previously at this stage? Ah, yes. Divide the first equation by the second. Doing this gives

$$\sqrt{2} = \frac{m}{n}$$

which is not possible since the Ancient Greeks proved that $\sqrt{2}$ cannot be written as a fraction. Consequently, either the tiles in the topmost row or rightmost column (or both) would have to be cut in order to fill the rectangle exactly. If all the tiles are left perfectly square, *no matter what their size*, they will not exactly fit the rectangle.

Bravo!

On Parade

Does $\sqrt{2}$'s irrationality have any other consequences? I'm fully hooked at this stage.

I would like to expand a little in a visual manner on the fact that twice the square of a natural number is never the square of another natural number.

"Twice a square is never a square"? Of course, this just means that $m^2 = 2n^2$ is not possible. I, of all people, should know this, given the time I spent trying to find a match between one of the first thirty perfect squares

$$1, \quad 4, \quad 9, \quad 16, \quad 25, \ldots, 729, \quad 784, \quad 841, \quad 900$$

and one of their doubles. Twice the square of n is $2n^2$, and it can never be equal to another square such as m^2.

Where m and n are positive integers. Because if there were an instance of this happening, then $\sqrt{2}$ would be rational.

Understood.

If you were a drill sergeant in command of a perfect square number of soldiers, then you could parade your squadron in a square formation.

Okay, a bit of exercise would be welcome.

As I was saying, with each rank and file having exactly the same number of soldiers, who, being properly trained by you to keep the same distance from all of their neighbours at all times, would march together in the shape of a geometric square.

In perfect formation. I'd expect nothing less.

That's the idea. Five abreast and five deep

to form a five-by-five square. Now, drill sergeant, I'm going to double your squadron to fifty. What do you make of that?

I'm much more important now?

Maybe, but really you're very sad because you can no longer parade this enlarged squadron in the square formation that you find so pleasing.

I am? I think you're trying to make it look like I have a screw loose!

No, you're an earnest sergeant who likes things just perfect; and what's more perfect from a drill sergeant's point of view than a square formation?

Okay, so I'm a crazy drill sergeant.

Now what if I assign you a different perfect square number of soldiers initially and then come along and double your squadron afterwards?

> The perfect square numbers, again, are: 1,4,9,16, . . . , right?

Yes, the squares of the natural numbers.

> Well, you'll drive me even crazier. In the beginning I'll be content because no matter which of the perfect squares you choose, I can have them parade in the shape of a square, which I'm supposed to love so much. But when you double this number of soldiers, I get cranky because I can no longer parade them in a square formation.

Why?

> Because the proof of the irrationality of $\sqrt{2}$ shows that twice a square is never a square. So, we might say, that here we have another consequence of the irrationality of $\sqrt{2}$?

Yes, or perhaps a consequence of the proof of the irrationality of $\sqrt{2}$.

> I notice that the number 50, which is the size of my original enlarged squadron, is nearly a perfect square. If I make 1 of the 50 stand out in front with a flag or a bugle or something, then I can arrange the remaining 49 into a 7×7 formation behind this leading soldier:

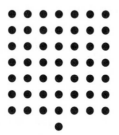

You can indeed. A configuration with a certain symmetry to it.

> So maybe I can cheer myself up a little bit when you come around doubling up the number of soldiers I must parade.

Maybe. With this particular setup for the 50 soldiers, I suppose we could say that you achieve the next best thing to the unattainable ideal of a perfect square formation.

> Okay.

But what if, instead of having an extra soldier, you are short a soldier?

This is pretty well the same, maybe even better because then I could make up for the missing person by filling the vacant position myself to complete the square.

And lead from the front.

Of course, since I'm in charge.

So, for the sake of impartiality, we could say that one soldier above or below a perfect square number of soldiers is the next best thing to the ideal because they can be paraded in a near-perfect square formation, to coin a phrase.

Not a bad description. It has just struck me, since $50 = 2(5)^2 = 7^2 + 1$, that this talk about near-perfect square formations ties in with our previous notion of near-misses.

Those perfect square numbers that are either one more or one less than twice another perfect square number?

Yes. The parading of a doubled squadron that is one soldier in excess or short of a perfect square in a near-perfect square formation is just a way of visualizing the numerical relationship.

You are absolutely correct. Each is the other in a different guise.

This means that I already know quite a few squadrons that can parade in the shape of a square and that, if doubled, can march in a near-perfect square formation.

Remind us how to find their sizes.

Just consult the terms of the sequence

$$\frac{1}{1}, \frac{3}{2}, \frac{7}{5}, \frac{17}{12}, \frac{41}{29}, \frac{99}{70}, \frac{239}{169}, \frac{577}{408}, \dots$$

and the sizes we seek are all there before our eyes.

Where, for example?

Well, the third fraction, $\frac{7}{5}$, is directly related to the two squads we have been discussing. Its denominator 5 is the number of soldiers in each rank or file of the smaller squadron with its $25 = 5^2$ – a perfect square of soldiers, while its numerator 7 gives the number in each rank or file of the larger squadron, where either I am filling out the final row or where there's one soldier out in front.

And the fact that $7^2 - 2(5)^2 = -1$, which is the same as $2(5)^2 = 7^2 + 1$, shows why this is possible?

Yes. Had we not known it already, the second equation tells us that if we choose the original squadron to have 5^2 soldiers, then double this squadron has 7^2 soldiers plus 1 and so can be paraded in a near-perfect square formation.

It would.

Since the square of the numerator of each fraction in the sequence is twice its denominator squared plus or minus 1, we know a whole host of squadrons with a perfect square number of soldiers, which when doubled in number, can be paraded in near-perfect square formations.

Remember that we haven't proved that this is true for *every* term in the sequence.

I know, but we do know it's true for the fractions displayed.

Granted, because we verified it in every single case.

And we can calculate from these fractions whether double the number of the squad is 1 short or 1 in excess. The calculation $1^2 - 2(1)^2 = -1$ associated with the fraction $\frac{1}{1}$, for example, says that $2(1)^2 = 1^2 + 1$, and so points to an excess soldier.

Thus, if you start with one lone soldier making up both the single rank and the single file of the trivial square:

●

then double this number of soldiers can be arranged into a near-perfect square formation, where one soldier is out in front of the square behind him.

Yes, like so:

●
●

This arrangement that, though it may not look like it, is a near-perfect square formation as it is just one over a perfect square.

Yes.

With such small numbers it can be hard to discern patterns.

For the fraction $\frac{3}{2}$, we have that $3^2 - 2(2)^2 = 1$ or, equivalently, that $2(2)^2 = 3^2 - 1$. This tells us that if we double an original squadron of $2^2 = 4$ soldiers, which can be arranged in a square with two rows and two columns:

● ●
● ●

then the enlarged squadron will need one soldier to complete a square formation.

We can see that this is the case when the eight soldiers are arranged thus:

And as already mentioned, if the drill sergeant falls in at the middle of the front row, we obtain a square, so this formation is a near-perfect square.

The next fraction in the sequence is $\frac{7}{5}$, and we have seen how this works.

Since there is no fraction in the sequence with a denominator of 3, it might be a nice exercise to take an initial squadron of size $3^2 = 9$, which can be arranged in this square:

and examine whether double this number of soldiers can be arranged to form a near-perfect square formation.

A good idea. Now $18 = 16 + 2$ tells us straight away that it is not possible.

You are right, of course. We could settle for a formation such as:

which has its own charms but does not satisfy our requirements.

Is it going to turn out that the smaller squadron sizes must be the squares of the following numbers

$$1, \quad 2, \quad 5, \quad 12, \quad 29, \quad 70, \quad 169, \quad 408, \ldots$$

and no other numbers?

It could well be.

> These numbers are just the denominators of the fractions in our previous sequence.

Yes, I can see that. You might like to know that this sequence is known as the Pell sequence.

John Pell
(1611–1685)

> It must be an important sequence if it has got a name.

Well, if what you think about these numbers is true, then it is an important sequence. You might like to puzzle out how to generate successive terms in it.

> I know how to get more and more terms in it because I know how the terms in our fraction sequence are generated.

You do indeed. But can you find a rule that just relates to the terms in this sequence alone?

> You mean some rule that refers only to the sequence itself and doesn't use outside help, as it were?

Exactly.

> You must give me lots of time.

Take all the time you need. You understand that what you speculate in relation to the Pell sequence is a big generalization from what we have established as fact.

> I realize this. I would like you to tell me if I am right, but I would also like you to teach me how to prove for myself the observations we have made.

Ambitious, and very laudable. Maybe none of what you believe to be true is so in general.

> I'll be most surprised if what we have observed to date is false in general, as you say.

Before we set about this mission, I would like to discuss one further consequence of the irrationality of $\sqrt{2}$ as it relates to the nature of its decimal expansion.

> Should be interesting.

The Nature of the Decimal Expansion of $\sqrt{2}$

The irrationality of $\sqrt{2}$ places certain restrictions on its possible decimal expansion.

> It does?

Yes. For starters, its decimal expansion cannot be like

$$1.40000000000000000000\ldots$$

where the 4 after the decimal point is followed by an endless string of zeros.

Isn't an expansion such as this one normally written without all those zeros at the end?

Yes. It is simply written as 1.4.

That's what I thought.

A decimal expansion that terminates in all zeros is called a *terminating decimal expansion,* or a *terminating decimal* for short.

And $\sqrt{2}$ does not have a terminating decimal expansion, you say?

That's correct. Do you think you could explain *why* it can't have such an expansion?

I think so. Because the decimal 1.4 has just a single digit appearing after the decimal point, I multiply it by the fraction $\frac{10}{10}$ to get that

$$1.4 \times \frac{10}{10} = \frac{14}{10} = \frac{7}{5}$$

So 1.4 corresponds to the fraction $\frac{7}{5}$, thus it is not equal to $\sqrt{2}$.

Good. Since $\frac{10}{10}$ is the same as 1, multiplying by it is just a clever device to write 1.4 as a fraction.

Of course, I wouldn't normally be as long-winded as this calculation makes me out to be.

Meaning?

I'd just write

$$1.4 = \frac{14}{10}$$

straight off and then cancel the factor 2 common to the numerator and denominator. Anyway, to answer your question, I can do pretty much the same for any terminating decimal: show that it is just another way of writing a fraction, and so $\sqrt{2}$ couldn't have a terminating decimal expansion.

Show us what you would do for the terminating decimal 0.152.

Because it has three digits after the decimal point, I multiply it by $\frac{1000}{1000}$.

The fraction with 10 to the power of 3 for both its numerator and denominator. So you multiply it "above and below" as is often said, by a power of 10 which matches the number of digits behind the decimal point.

I suppose so.

Since $0.152 \times 10^3 = 0.152 \times 1000 = 152$, you simply write that

$$0.152 = \frac{152}{1000}$$

Then you see if you can reduce the fraction to its lowest form by canceling factors common to its numerator and denominator.

> Yes. Now that you mention it, I suppose I should try. I see that I can divide above and below by 8 to get

$$0.152 = \frac{19}{125}$$

> Since the numerator of this fraction is the prime number 19, I know that I cannot reduce it any further.

Nicely spotted. So you have written the decimal 0.152 as a fraction in lowest form. I suppose you can verify that $\frac{19}{125}$ has the decimal expansion 0.152 using long division.

$$\begin{array}{r} 0.152 \\ 125\overline{)19000} \\ \underline{125} \\ 650 \\ \underline{625} \\ 250 \\ \underline{250} \\ 000 \end{array}$$

> I think I could but, as I said before, I wouldn't like to have to explain the procedure, though I'm sure it's not that hard to explain.

Nor will you have to. Anyway, by imitating what you have done with the two terminating decimals 1.4 and 0.152, we can show that every terminating decimal expansion represents a fraction. And since $\sqrt{2}$ is irrational, its decimal expansion cannot be a terminating one.

> So that settles that.

But there is more to this story than merely saying that $\sqrt{2}$ cannot have a terminating decimal expansion.

> There is?

It cannot have a decimal expansion such as

$$0.62428571428571428571\ldots$$

either. Here the initial two digits after the decimal point, 62, are followed by the six-digit block 428571, which in turn is followed by another block with the six digits 428571, in the exact same order, and so on, with this six-digit block repeating itself indefinitely.

> Let me examine this closely. You are saying that after the digits 6 and 2, there is a pattern consisting of blocks of 428571 repeating itself over and over again.

Ad infinitum, or all the way to infinity.

> Okay, I believe this.

Now $\sqrt{2}$ cannot have a decimal expansion such as this one either, where after a certain number of initial digits, a block of digits repeats itself over and over again. A decimal expansion of this type is said to be *mixed*.

> So what kind of number has this type of decimal expansion? Is it very complicated?

Now you have posed yourself a little puzzle.

> I have? I asked you the question fully expecting you to enlighten me, but now I sense a task coming on.

Of course, maybe you could experiment a little. Perhaps the trick you used to find the fraction corresponding to a terminating decimal could help in some way.

> But those two decimal expansions are finite. This expansion is much more intimidating because it's infinite even if after a while it is only the same block of six digits repeated over and over.

Something we'll exploit nicely, as you'll soon see.

> You'll have to show me.

I will, but before I do I want you to tell me how I might get those two digits, 62, immediately following the decimal point in

$$0.62428571428571428571\ldots$$

out in front of the decimal point so that what then follows the decimal point is nothing more than the six-digit block 428571 repeated indefinitely.

> Just use the previous trick of multiplying by $10^2 = 100$. You'll get a number where the first two digits, 62, appear in front of the decimal point, with the block 428571 repeated over and over behind the decimal point.

So

$$(0.62428571428571428571\ldots) \times 100 = 62.428571428571428571\ldots$$

> Yes, that looks right.

Now, if we subtract 62 from this number, we get

$$0.428571428571428571\ldots$$

which is a purer specimen. I say this because in a sense it's as if the impurities have been removed from the original expansion to give one that consists of an infinite number of copies of the first block of six digits appearing immediately after the decimal point.

> So the decimal expansion $0.428571428571428571\ldots$ is not a mixed one?

That's right. The expansion we have now is said to be *purely periodic*. Pure because there are no digits following the decimal point that are not part of the repeating pattern, and periodic precisely because of the infinite repeating of the six-digit block 428571. This periodic decimal expansion is said to have "period 6" because it's a six-digit block that repeats itself.

> Sensible.

Because of this periodicity, the infinite decimal expansion is sometimes written more succinctly as

$$0.\overline{428571}$$

—it being understood that the overlined block is to be repeated indefinitely.

> So $0.\overline{3}$ stands for 0.3333333333333333 ... ?

Yes.

> I remember this from school. This is the decimal expansion of the fraction $\frac{1}{3}$ if I am not mistaken.

You are not. A very nice way of showing that this is so is to set

$$x = 0.3333333333333333\ldots$$

Then multiply by 10 to get

$$10x = 3.3333333333333333\ldots$$

Then subtract the original expansion from this one:

$$10x = 3.3333333333333333\ldots$$
$$\underline{x = 0.3333333333333333\ldots}$$
$$9x = 3.00000000000000000\ldots$$

to get

$$9x = 3 \Rightarrow x = \frac{3}{9} = \frac{1}{3}$$

Then we have your fraction.

> That's ingenious. The way the 3's behind the decimal points in both expansions are made to "kill each other off," so that we end up with a terminating decimal.

Simple but clever.

> Who spots these tricks?

I have often asked myself the same question about these inspired flashes of insight. But coming back to the decimal expansion 0.62428571428571428571 ... , which in our new notation we could write as $0.62\overline{428571}$, why don't you try finding out the value of $0.\overline{428571}$ using this technique?

I'll give it a go. So I begin by writing

$$x = 0.428571428571428571\ldots$$

using the symbol x again.

A new job for x, which is always called on whenever there's something to be found out.

A very busy fellow! This problem is a little more challenging than the one you just did, but I think I see what to do.

Great.

Since the period of this expansion is 6, I'll multiply by 10^6 to get

$$1000000x = 428571.428571428571428571\ldots$$

and then subtract x from it.

That's all there is to it.

So we get

$$
\begin{array}{r}
1000000x = 428571.428571428571428571\ldots \\
x = 0.428571428571428571\ldots \\
\hline
999999x = 428571.00000000000000000\ldots
\end{array}
$$

Then

$$999999x = 428571 \Rightarrow x = \frac{428571}{999999}$$

and so

$$0.428571428571428571\ldots = \frac{428571}{999999}$$

We have ended up with a fraction.

You have indeed.

Is it in lowest possible form?

Probably not. It would be nice to see it in lowest terms, but it is not necessary and does involve a lot of labor.

But I'm most curious so see if it boils down to being a simple fraction.

You may like to show that

$$0.428571428571428571\ldots = \frac{428571}{999999} = \frac{3}{7}$$

It shouldn't take all that long.

> Who would have thought this decimal expansion is that of the innocent-looking fraction $\frac{3}{7}$?

The problem of reducing fractions involves factorization, which can be far from easy, particularly if the numbers are really large.

> Still, it's clever how the infinite is turned into the finite.

A mathematical poet. You might like to test your mettle on taming the expansion

$$0.\overline{012345679}$$

which has, with the exception of 8, all the digits appearing in their usual order.

> Intriguing. I have to see which fraction has this wonderful purely periodic expansion.

And you are convinced that you'll end up with a fraction?

> I fully expect to, because when I perform exactly the same tricks as above I'll end up with

$$999999999x = 12345679 \Rightarrow x = \frac{12345679}{999999999}$$

> showing that $x = 0.\overline{012345679}$ is a fraction.

Excellent. Now enjoy yourself reducing this faction to it lowest form. You'll get a surprise.

> I'd love to do it now, but I suppose we had better finish what we were doing.

Yes, we had better remind ourselves what it is we have just achieved and what we are about.

> We have shown that

$$0.\overline{428571} = \frac{3}{7}$$

> or if you prefer the "big screen" version, that

$$0.42857142857142\ldots = \frac{3}{7}$$

Our original task was to find the number represented by

$$0.6242857142857142\ldots$$

with the "impurity 0.62 at the front of it." Well, everything is ready, so I think you should now complete this in fairly short order.

Let me check back then. We showed that

$$(0.6242857142857142\ldots) \times 100 = 62.42857142857142\ldots$$

and then we showed that $0.\overline{428571}$ is $\frac{3}{7}$. So we may write that

$$100(0.6242857142857142\ldots) = 62.42857142857142857142\ldots$$

$$= 62 + 0.42857142857142857142\ldots$$

$$= 62 + \frac{3}{7}$$

$$= \frac{(62 \times 7) + 3}{7}$$

$$\Rightarrow 100(0.6242857142857142\ldots) = \frac{437}{7}$$

$$\Rightarrow 0.6242857142857142\ldots = \frac{437}{700}$$

At last!

Very nicely done. This result can be verified by long division or a calculator.

So the mixed decimal

$$0.62\overline{428571} = \frac{437}{700}$$

also turns out to be that of a fraction.

I think that it is clear from what we have just done that *any* decimal expansion consisting of a finite number of digits after its decimal point, followed by a finite block of digits that repeats itself endlessly, represents a rational number.

I can see that this is true because there is nothing to stop us from imitating exactly the steps taken in the above argument to produce a rational number every time.

That's right. So what implications has all this for the decimal expansion of our irrational friend $\sqrt{2}$?

Ah yes! It implies that its decimal expansion is not periodic from some digit onwards.

Because?

Because if it were, then $\sqrt{2}$ would be a rational number.

Something we know it definitely is not.

It is interesting that, without actually knowing the nature of the decimal expansion of $\sqrt{2}$, we still can say that it is neither terminating nor periodic from some stage on.

Negative things to say, in a way, but facts nonetheless.

I can believe from the particular examples we have discussed that every terminating decimal, mixed decimal, and purely periodic decimal expansion represents a fraction, but does *every* rational number have an expansion which is one or other of these three types?

A good question, to which the answer is "Yes." If we were to take the time to examine carefully the long-division procedure for obtaining the decimal expansion of a fraction, we would understand that such expansions *must* either terminate, be mixed, or purely periodic.

So if I examine a decimal expansion and find it has no repetitive pattern, then I know it is the decimal expansion of an irrational number.

I'm going to say yes, although no human being or computer could ever hope to examine a complete decimal expansion so as to be able to pronounce that a block of digits doesn't start repeating from some point onwards.

Okay, if I examine the first million digits of a decimal expansion very thoroughly and do not spot a repetitive structure, I still cannot say anything about the rationality or irrationality of the number represented by the complete expansion?

I'm afraid not. Even though a million digits is a sizeable number of digits, you might be looking at only a small part of the leading nonperiodic portion of a mixed decimal.

With such a length?

Yes, any length you like. Or you could be looking at a mixture, where you see all of the nonperiodic portion of the expansion and some of its periodic part.

But not enough to know that the expansion had entered a repetitive cycle?

Something like that.

And could I be looking at a purely periodic expansion and not know it?

Easily, if the period of the decimal is more than a million. The fraction

$$\frac{1}{98982277}$$

has a period of 16,493,730.

> Simply staggering! And is it hard to generate the successive digits of this decimal expansion?

Not at all for this or any rational number. It takes no more than a line of computer code to simulate the long-division algorithm. The decimal expansion of this fraction begins after the decimal point with the digits

$$0, \quad 0, \quad 0, \quad 0, \quad 0, \quad 0, \quad 0, \quad 1, \quad 0, \quad 1, \quad 0, \quad 2, \quad 8$$

which are then followed by a sequence of digits that have all the appearance of being completely random. Then, after 16,493,730 digits, the same thirteen digits

$$0, \quad 0, \quad 0, \quad 0, \quad 0, \quad 0, \quad 0, \quad 1, \quad 0, \quad 1, \quad 0, \quad 2, \quad 8$$

reappear, followed by exactly the same sequence of digits that came after these thirteen digits on the first cycle.

> Absolutely fascinating!

Yes, the study of periodic decimal expansions is captivating and holds many gems. The decimal expansion of $\frac{1}{61}$ is

0.016393442622950819672131147540983606557377049180327868852459

Here each digit appears six times after the decimal point. The relative frequency of each digit is exactly the same for each digit! Check it out.

> Amazing!

You could use this expansion to assign six tasks each to ten people in an apparently random way.

> I'll be sure to use this knowledge when the need arises! Are there many fractions of this type where each digit occurs exactly the same number of times in the decimal expansion?

I don't know, although I know how to fish for them and have landed a couple of beauties. Let this question be an investigation for yourself.

> For a future time. I think I understand the real problem with the question I asked.

Which is?

> In this case, that one cannot say what the nature of a number represented by an infinite decimal expansion is from an examination of only a finite amount of that expansion. To draw con-

clusions we'd have to make assumptions about the vast unseen portion . . .

. . . which lies hidden in "the fog of infinity," to borrow a metaphor from a medieval mathematician. However, you can use what we know about the decimal expansions of rational numbers to construct irrational numbers.

How?

Let me give you an example. Take all the natural numbers

1, 2, 3, 4, 5, 6, 7, 8, 9, 10, 11, 12, 13, 14,
15, 16, 17, 18, 19, 20, 21 . . .

and use them to construct the decimal expansion

$$0.12345678910111213141516171819202 1 \ldots$$

The base-ten Champernowne constant.

In this expansion, each natural number appears in its usual order after the decimal point.

This is done deliberately?

To ensure that the decimal expansion cannot possibly have a repetitive structure beginning anywhere along the entire expansion.

Which means that the expansion cannot be that of a rational number. Very clever.

Isn't it? If we accept that the decimal expansion formed in this manner could never settle down to being periodic, then it must be the expansion of an irrational number.

It seems almost obvious that it couldn't have a repeating pattern from some point on.

You might like to turn "almost obvious" into "completely obvious" in an idle moment.

If I ever get any! Do you have any other examples?

I have, but we shouldn't let ourselves stray too much.

Just one more, then.

The infinite decimal expansion

$$0.10100100000010000000000000000000000000100 \ldots$$

is that of another irrational number.

Built from just zeros and ones?

But do you see how?

Something else to think about during those idle moments. Which irrational numbers do these two cleverly constructed expansions represent?

By which you mean, are they the decimal expansions of irra-tional numbers such as $\sqrt{2}$?

I suppose that's what I mean.

I have no idea if the first one represents any irrational quantity of this type, or of any other type for that matter, that can be connected to rational numbers. I'm told on good authority that the second one is not of the "$\sqrt{2}$ type," if I may say this loosely.

An even stranger number, then?

You could say that. It is best to think of these numbers as be-ing solely defined by their expansions. We could be diverted forever if we were to begin probing the nature of this breed of irrational number.

Well, this discussion of decimal expansions has given me a better understanding of how they relate to rational and irra-tional numbers. I'm really interested in seeing more of the digits of the expansion of $\sqrt{2}$, now that I know that its irrational nature prevents its decimal expansion from having a periodic structure from some point onward.

That is good to hear. I think it is time we concluded this excur-sion describing some of the consequences of the irrationality of $\sqrt{2}$.

Before you do, I have a question that I meant to ask some time back, when we were talking about squeezing $\sqrt{2}$ into smaller and smaller intervals.

Which is?

Can't we get as many places of the decimal expansion of $\sqrt{2}$ as we want simply by narrowing in on it by tenths?

Will you elaborate, please?

Well, when we show that

$$(1.4)^2 = 1.96 \quad \text{and} \quad (1.5)^2 = 2.25$$

we know that

$$1.4 < \sqrt{2} < 1.5$$

Yes, we know the location of $\sqrt{2}$ on the number line to within one-tenth of a unit because we know it lies in the interval [1.4, 1.5].

So now can't we find in which tenth of this interval it is?

Yes, this interval itself can be subdivided into the ten subintervals

$$[1.40, 1.41], \quad [1.41, 1.42], \ldots, [1.48, 1.49], \quad [1.49, 1.50]$$

and we can determine which of these subintervals $\sqrt{2}$ lies in.

My point exactly. In fact, since $(1.4)^2 = 1.96$, and $(1.5)^2 = 2.25$, we know that $\sqrt{2}$ is much closer to 1.4 than it is to 1.5.

So?

Well, it means I'd check the subintervals starting with [1.40, 1.41] first.

Agreed.

I already know that $(1.40)^2 = 1.96$, and I can work out by hand that $(1.41)^2 = 1.9881$. Since this is still less than 2, I now work out $(1.42)^2$ to get 2.0164. At this stage, I know that

$$1.41 < \sqrt{2} < 1.42$$

Agreed. So you now know the decimal expansion of $\sqrt{2}$ to one decimal place.

By the way, can you tell me why $\sqrt{2}$ will never land on the endpoint of a subinterval?

Because if it did it would have a terminating decimal expansion, which we know it hasn't.

Excellent. What are you going to do now?

Subdivide the interval [1.41, 1.42] into the ten subintervals:

[1.410, 1.411],   [1.411, 1.412],   . . .   [1.418, 1.419],
    [1.419, 1.420]

and determine which of these subintervals $\sqrt{2}$ lies in.

Where are you going to start?

I'd have to think about this. Probably by squaring 1.411.

Well, if you go about it this way, you'll work out that

$$(1.411)^2 = 1.990921$$
$$(1.412)^2 = 1.993744$$
$$(1.413)^2 = 1.996569$$
$$(1.414)^2 = 1.999396$$
$$(1.415)^2 = 2.002225$$

to find that

$$1.414 < \sqrt{2} < 1.415$$

I might have skipped some of these squarings and gone straight to $(1.414)^2$.

Even if you did, you can see that there is a lot of work involved in testing the squares. You now know $\sqrt{2}$ to two decimal places.

I'm beginning to see the light. There will be a lot more work involved to get the third decimal place, and even more to get the fourth and so on.

This process, repeated over and over again, defines what is meant by a decimal expansion. In theory, it allows us to determine as many of the leading digits of a number's decimal expansion as we may want.

In theory maybe, but very slow in practice?

Yes, when you compare this tedious method with what we already know through your rule. With no more than a small number of simple additions and a few elementary multiplications and divisions, we found that

$$\frac{239}{169} < \sqrt{2} < \frac{577}{408}$$

Now the long-division algorithm applied to the two fractions, which does not take too long in these two cases, gives

$$1.414201183431952\ldots < \sqrt{2} < 1.4142156862745099\ldots$$

... the first four decimal digits of the decimal expansion—impressive. Generating more and more fractions would seem like a better way to go.

Perhaps, but provided we can settle a number of questions that we left unanswered.

CHAPTER 3

The Power of a Little Algebra

So we are going to study the sequence

$$\frac{1}{1}, \frac{3}{2}, \frac{7}{5}, \frac{17}{12}, \frac{41}{29}, \frac{99}{70}, \frac{239}{169}, \frac{577}{408}, \ldots$$

in earnest?

Definitely. The waiting is over.

The first thing I want you to prove is the plus or minus 1 property.

Oh yes, your conjecture that *every* term in this sequence has the property that its "numerator squared minus twice its denominator squared" alternates between −1 and 1.

Or maybe you would show me how to prove it for myself.

But this property might not be true in general.

I'll eat my hat if it isn't!

I know you have verified that it holds for the first eight fractions, but couldn't all of this be just a fluke, circumstantial evidence, as the lawyers might say?

Would it not have to be one massive fluke?

Perhaps, but maybe it just is. It might be that none of the fractions, as yet ungenerated, has this property, or it could be that some will have it but others won't.

Logically, I know that what you say is correct until we can show otherwise, but I'm sticking to my hunch that every fraction in the sequence has this property.

Such conviction! Right, then, let our first job be to learn if this alternating property propagates itself along the entire infinite sequence.

Alternating property? Nice—a shorter way of describing the plus or minus 1 property.

You realize that such a proof, if there is one, must involve some algebra.

By this stage I think I appreciate that if an infinite number of cases have to be considered, then algebra must be used.

Yes, whereas arithmetic can be used to check a finite number of cases, it is unable to cope when the number of possibilities is infinite.

Simply because they cannot all be examined individually; which is the problem with checking this alternating property.

Yes. It is algebra that enables us to prove general truths. One of its strong selling points, you could say.

Which it badly needs for people like me, who are inclined to tremble at the thought of being expected to use any.

It's no secret that many people simply turn off at its mere mention; but you've been doing very well so far.

Well, I think so. It is easier to be more positive about its use when you understand why it is so necessary, and easier still when you see what it can achieve.

Let's begin on that upbeat note.

But where should we start?

Why don't we remind ourselves how the sequence

$$\frac{1}{1}, \ \frac{3}{2}, \ \frac{7}{5}, \ \frac{17}{12}, \ \frac{41}{29}, \ \frac{99}{70}, \ \dots$$

is generated by the wonderful rule that you discovered earlier.

This rule says:

To get the denominator of the next fraction in the sequence, add the numerator and denominator of the previous fraction. To get the numerator of the next fraction in the sequence, add the numerator of the previous fraction to twice its denominator.

As we said before, all fairly straightforward.

Almost surprisingly so. Now can we translate this somewhat lengthy verbal description into a shorter but easily understood mathematical rule.

No doubt this is where letters will come in handy.

Yes. If we find a good way of doing this, we'll almost surely reap rich rewards.

Sounds promising.

To make what I am going to do next a little easier, I hope, I'm going to make a small change in the wording of the rule—it won't change the result. I'll use the word *current* instead of the word *previous* so that the rule now reads:

> To get the denominator of the next fraction in the sequence, add the numerator and denominator of the current fraction. To get the numerator of the next fraction in the sequence, add the numerator of the current fraction to twice its denominator.

So in terms of generating successive terms of the sequence using this rule, the current fraction is the one we have before us at present?

Or the one just generated a moment ago and on which we are about to apply the rule so as to generate the next fraction in the sequence. How might we describe this current term using letters?

By

$$\frac{m}{n}$$

as we have done before?

Why not? With m standing for the numerator and n for the denominator. With this notation we are not committed to any specific fraction, thus we have the freedom to talk about any of the fractions in the sequence without naming any one numerically. Should we want to talk about a particular fraction, we give the general numerator m and the general denominator n specific numerical values, which then identify the fraction in question.

So if we want to talk about the third term in the sequence, $\frac{7}{5}$, we say m has the value 7 and n the value 5 in this specific case.

Exactly.

And if $\frac{m}{n}$ is the sixth fraction $\frac{99}{70}$, then $m = 99$ and $n = 70$.

Why don't you begin translating your descriptive rule into one involving m's and n's.

I'll try. With regard to denominators, the recipe says that to get the denominator of the next term, add the numerator and denominator of the current fraction.

Correct.

So "add the numerator and the denominator of the current fraction" translates to $m + n$.

Exactly. Nothing more than a simple $m + n$. You might say that algebra is nothing more than arithmetic applied to letters.

That wasn't so hard.

Nothing mysterious whatsoever, once you get the hang of it. Now do the same for the numerator of the next fraction.

> Okay. The rule says that to get the next numerator, "add the numerator of the current fraction to twice its denominator." This translates to $m + 2n$.

Again, very straightforward.

> Yes, once you have been guided.

So what is the next fraction after $\frac{m}{n}$ in the sequence?

> According to what we have just said, I suppose it's
>
> $$\frac{m+2n}{m+n}$$
>
> which is not all that terrifying.

Good. You have done a great job in writing the descriptive rule in a more compact way. Sometimes we write

$$\frac{m}{n} \to \frac{m+2n}{m+n}$$

and read the $\to$ as "becomes."

> So in words, the rule says that "m over n becomes $m + 2n$ over $m + n$," is that it?

That's one way of saying it. We often use phrases such as "transforms into" or "maps into" or "generates" in place of "becomes." In any event the rule describes how the typical fraction $\frac{m}{n}$ is transformed or carried into the next fraction $\frac{m+2n}{m+n}$. Even though it contains terse symbols rather than familiar words, this expression of the rule does have the virtue of being much shorter and so much easier to write down.

> I'd have to agree, but it may take some getting used to.

Why don't you check out the "algebraic" form of the rule on a specific case?

> All right, it will give me practice. Which fraction do you recommend I apply it to?

First off, why not test it on the starting fraction, $\frac{1}{1}$?

> In this case, $m = 1$ and $n = 1$, so the above expression reads
>
> $$\frac{1}{1} \to \frac{1+2(1)}{1+1} = \frac{3}{2}$$

Correct.

Now try it on the result of this calculation.

> You mean on $\frac{3}{2}$?

Yes.

With $m = 3$ and $n = 2$, the algebraic rule gives

$$\frac{3}{2} \to \frac{3+2(2)}{3+2} = \frac{7}{5}$$

Correct again. You could now give this rule to a computer program, along with the starting fraction $\frac{1}{1}$, and the computer would generate hundreds of successive terms of the sequence in the twinkling of an eye, simply by doing what you have done.

No doubt at lightning speed. So we are making progress.

Most definitely.

Seed, Breed, and Generation

But what now?

Let us remind ourselves of what our mission is.

To show for every fraction in the sequence

$$\frac{1}{1}, \frac{3}{2}, \frac{7}{5}, \frac{17}{12}, \frac{41}{29}, \frac{99}{70}, \frac{239}{169}, \frac{577}{408}, \ldots$$

that the quantity "numerator squared minus twice its denominator squared" is either −1 or 1.

Exactly, to establish the alternating property. If successful, it will mean that we have an infinite source of perfect squares that are within 1 of being double another perfect square.

In the squares of all the numerators, the "near misses," as I called them.

None other, an infinity of near misses. To resume our discussion, we now know that this sequence is generated by applying the rule

$$\frac{m}{n} \to \frac{m+2n}{m+n}$$

first to the fraction $\frac{1}{1}$ to obtain the next term. Then the rule applied to this "newborn" fraction produces the next member of the sequence and so on, with the rule acting at each successive stage on the latest arrival.

Looked at in this way, it is only the seed and the rule that matter; everything else is predetermined.

I'm delighted you have made this observation. The fraction $\frac{1}{1}$ can be thought of as the "seed" that breeds the successive generations of the sequence by constant application of the same rule.

> And the sequence is a family tree with only one line of descent.

You could say that, and, viewed this way, what we hope to prove is that a certain characteristic of the original seed is passed from one generation to another.

> And would applying the same rule to a different seed generate a different sequence, with some original characteristic of its seed being preserved?

We can certainly say that the same rule applied originally to a different seed will produce different fractions at each generation stage. That the original characteristic would be propagated as well would have to be proven, but we are getting ahead of ourselves. For now, we want to show that

> The square of the numerator of any term minus twice the square of its denominator is either -1 or 1.

is true for every term of the sequence generated by the rule.

> The alternating property.

Shorter still might be, "Top squared minus twice bottom squared is either -1 or 1," using plain but less mathematical language.

> Is this acceptable?

Anything that helps us to think more clearly is to be welcomed.

> So "top squared minus twice bottom squared" it is.

See if you can translate this into a statement concerning m's and n's.

> Well, I suppose, since $\frac{m}{n}$ stands for *any* term in the sequence, the top squared is m^2.

True.

> And twice the square of the bottom is $2n^2$.

Right again.

> So $m^2 - 2n^2$ is just another way of saying "top squared minus twice bottom squared" in terms of m and n for the typical fraction $\frac{m}{n}$.

You are doing very well. Now then, what does the alternating property assert?

> That the quantity $m^2 - 2n^2$ alternates between -1 and 1 as $\frac{m}{n}$ goes from one term anywhere in the sequence to the next term.

That's it precisely. By the way, what is $m^2 - 2n^2$ when $\frac{m}{n}$ is the first term in the sequence?

Since $m = 1$ and $n = 1$ in this instance, $m^2 - 2n^2 = 1 - 2(1)^2 = 1 - 2 = -1$, which we knew already.

So the next time it should be 1, which it is, since $3^2 - 2(2)^2 = 9 - 8 = 1$, which we also knew already.

But how do I prove that this alternating pattern persists?

This is the crux of the matter. Express what you hope to achieve in terms of m and n.

Well, if $m^2 - 2n^2 = -1$ for the fraction $\frac{m}{n}$, then for the next fraction this quantity must be 1. And, vice versa, if $m^2 - 2n^2 = 1$ for the fraction $\frac{m}{n}$, then it must be -1 for the fraction after it.

What do you mean to say when you use the phrase "this quantity"? You don't mean $m^2 - 2n^2$ again, do you?

No. I realize that I'm being imprecise by expressing myself in this manner. What do I mean? I must pause to think.

Take your time.

I have it. I mean that the top of the *next* fraction squared minus twice the square of its bottom will be just the opposite of what it is for $\frac{m}{n}$.

Tremendous. Now you are getting things clear. In this problem we are not talking about the fraction $\frac{m}{n}$ alone, but also about the fraction following it.

This is the fraction

$$\frac{m+2n}{m+n}$$

according to the generation rule.

What you need to do now is translate "the top squared minus twice the bottom squared" into algebra for this fraction.

Oh, right. So now I must do for $\frac{m+2n}{m+n}$ what I did for $\frac{m}{n}$.

Correct.

In this case, the top squared is $(m + 2n)^2$ and the bottom squared is $(m + n)^2$. Will you work these out for me, as I wouldn't trust my school algebra?

Of course. First

[See chapter note 1.]

$$(m + 2n)^2 = m^2 + 4mn + 4n^2$$

while

$$(m + n)^2 = m^2 + 2mn + n^2$$

Thanks. I'll see how far I can push on now with this help. I get

(top squared)^2 − (twice bottom squared)^2 = $(m + 2n)^2 − 2(m + n)^2$

Now

$$(m + 2n)^2 − 2(m + n)^2 = (m^2 + 4mn + 4n^2) − 2(m^2 + 2mn + n^2)$$
$$= m^2 + 4mn + 4n^2 − 2m^2 − 4mn − 2n^2$$
$$= −m^2 + 2n^2$$

if I have done my calculations properly.

Flawlessly!

Considering that the fraction $\frac{m+2n}{m+n}$ looks more complicated than $\frac{m}{n}$, I would not have been surprised by a more complicated answer.

Nor would I have been, but the answer is surprisingly simple.

In fact, almost familiar. Doesn't the answer $−m^2 + 2n^2$ look very like the answer $m^2 − 2n^2$, obtained for the fraction $\frac{m}{n}$?

A crucial observation! Can you spot the exact connection between $−m^2 + 2n^2$ and $m^2 − 2n^2$?

Is one just the opposite of the other?

Yes, because $−m^2 + 2n^2 = −(m^2 − 2n^2)$. Now this relation is all you need in order to explain what we want to prove.

Let me see if I can explain why. We have shown that the top squared minus twice the bottom squared of the typical fraction $\frac{m}{n}$ is $m^2 − 2n^2$, while for the next fraction $\frac{m+2n}{m+n}$ this quantity is $−(m^2 − 2n^2)$.

Signifying ... ?

... whatever value the top squared minus twice the bottom squared has for $\frac{m}{n}$, it has exactly minus this value for the next fraction $\frac{m+2n}{m+n}$.

I cannot disagree. Please continue.

Since the top squared minus twice the bottom squared is actually $−1$ for $\frac{1}{1}$, the first fraction in the sequence, it must be 1 for the next fraction, $\frac{3}{2}$, then $−1$ again for the third term, $\frac{7}{5}$, and so on indefinitely.

Marvelous! You have explained algebraically why the numerator squared minus twice the denominator squared is always either $−1$ or 1 for every fraction in the sequence

$$\frac{1}{1}, \frac{3}{2}, \frac{7}{5}, \frac{17}{12}, \frac{41}{29}, \frac{99}{70}, \frac{239}{169}, \frac{577}{408}, \ldots$$

You have established the alternating property. Again, well done.

And is what we have done a proof?

It is, if we accept the following mathematical principle: imagine a ladder with equally spaced rungs stretching all the way to infinity. Mathematics accepts as a principle that if we can get on the first rung of this ladder and step from it to the next rung, then we can ascend the entire ladder.

But where's the ladder with its rungs here?

Here the ladder is the sequence of fractions, with the individual fractions being the equally spaced rungs.

So the seed is the first rung of the ladder.

Yes, and the step is showing that the property passes from this fraction to the next.

And so it steps along the whole sequence.

That's it.

It is fantastic to see the simple reason why the property that holds for the seed must propagate all along the sequence.

The power of a little algebra.

I could get to like the idea of being able to prove things using algebra. Understanding the reason why something is as it is makes one feel wiser. I want more challenges of this type.

Well, mathematics is just one never-ending sequence of challenges, some easy and some apparently insurmountable. Speaking of which, we have one challenge of our own still outstanding.

All-Inclusive or Not?

What is this challenge we are about to accept?

To answer a second question in connection with the sequence

$$\frac{1}{1}, \frac{3}{2}, \frac{7}{5}, \frac{17}{12}, \frac{41}{29}, \frac{99}{70}, \ldots$$

which you asked some time ago.

Which was?

How can we be sure that the fractions of this sequence are the only ones having the property that $m^2 - 2n^2$ is either plus or minus 1? As usual, $\frac{m}{n}$ stands for the typical fraction.

> I had forgotten, but I remember now that I was very interested in knowing the answer to this question when it first occurred to me.

Which is something I can well understand. If the alternating property is true only of this sequence, then this fact will be another measure of its uniqueness.

> But how are you going to go about answering this question?

The only way I know how: by turning the question over in my mind in the hope that some plan for tackling the problem will dawn on me.

> So just kicking it around until you see some way of getting started?

Yes, and even if this turns out to be fruitless, other ideas often strike you along the way. Now, to get started on this challenge, let me put the problem slightly differently: if someone shows me two integers, p and q, say, which are such that $p^2 - 2q^2 = 1$ or $p^2 - 2q^2 = -1$, may I then say that the fraction $\frac{p}{q}$ belongs to the above sequence, or could it be that it doesn't?

> Okay. By the way, you have used $\frac{m}{n}$ up to now to stand for a typical fraction. Is there a reason why you are changing to $\frac{p}{q}$?

We could still use $\frac{m}{n}$ or any other pair of letters, such as $\frac{n}{d}$ with n standing for numerator and d for denominator. Normally, using $\frac{p}{q}$ when you had been using $\frac{m}{n}$ would be viewed as no more than a change of clothing.

> It's not that I have any objection to $\frac{p}{q}$, but I wondered if there might be some significance in your choice of letters for this particular problem.

Only that in this case, it is preferable to use something other than $\frac{m}{n}$ precisely because up to now $\frac{m}{n}$ has stood for a typical member of the above sequence. Now we want to keep an open mind as to whether or not the fraction $\frac{p}{q}$ is a member of this sequence. By calling it something other than $\frac{m}{n}$, we avoid any such suggestion.

> I get the idea.

Have you any instincts as to the answer to this question of yours?

> Well, I must have thought that the fraction $\frac{p}{q}$ would have to belong to this sequence because I hadn't come across any excep-

tions during my searches among the first thirty squares and their doubles. However, I cannot think of any reason why this should always be the case.

So you have an open mind on the question.

Yes, I suppose. There could be maverick $\frac{p}{q}$'s.

Meaning fractions *not* in "our" sequence

$$\frac{1}{1}, \frac{3}{2}, \frac{7}{5}, \frac{17}{12}, \frac{41}{29}, \frac{99}{70}, \ldots$$

but having the property that their numerator squared minus twice their denominator squared is either plus or minus 1.

Yes.

Well then, I think the word "maverick" captures what you have in mind quite well. Let me give you the fraction

$$\frac{47321}{33461}$$

and show you that

$$47321^2 - 2(33461)^2 = 2239277041 - 2(1119638521) = -1$$

Is this fraction a maverick? What do you think?

That we haven't computed enough terms of our sequence for me to check if this specimen belongs to our sequence.

I understand that if you compute some more terms of the sequence and you come across this fraction, then you'll know for certain that it is not a maverick. In fact, it is the thirteenth term in the sequence. But how would you show that a particular maverick, if it exists, is not in the sequence?

Could we not use the fact that the number of digits in both the numerator and denominator of the fractions in the sequence gets bigger as we move out along the sequence?

Unfortunately this is only an observation; we've never actually proved it to be a general fact.

You are right, I know, but supposing it is true . . .

. . . for argument's sake?

Yes; then would we not simply need to generate enough fractions of the sequence until we get to one whose denominator is either that of the fraction we are testing . . .

. . . such as the denominator 33461 of a moment ago?

Yes, or until the denominator of the fraction being tested is passed by without having appeared as the denominator of the fractions being generated.

In which case you'd know that the fraction is a maverick.

> That's the idea: to generate enough terms of the sequence to show eventually that the fraction being tested is or is not in the sequence.

And if it were not in the sequence, then we'd have a maverick that would put an end to the whole matter.

> Certainly.

But even if we allow that this procedure is a valid one, which, by the way, could take an awful lot of time to implement on a fraction with a huge numerator and denominator, what happens if you don't find any maverick fraction?

> I realize that this is the real issue. If even after millions of tests we did not find a maverick, this wouldn't prove that they didn't exist in general, although it might make me believe very strongly that they didn't.

So have you any other ideas?

> To show that maverick fractions cannot exist by proving that if $p^2 - 2q^2$ is plus or minus 1 for some $\frac{p}{q}$, then this fraction lives somewhere in the sequence.

If we could prove this, then it would settle the matter also.

> But I haven't a clue how to go about proving this, if in fact it is true.

Well, is there anything else you could have done to check whether or not the fraction $\frac{47321}{33461}$ is in the sequence besides calculating more terms of the sequence explicitly to the point where this fraction made its appearance?

> Let me think. Maybe I see another way. If the fraction is in the sequence, then by working "backwards" I would hit upon a term that I recognise to be in the sequence.

And would this be enough?

> Surely, because by going forward from this known fraction I would come upon the fraction being tested.

I agree. So could you not adopt this strategy with every fraction to be tested?

> Hold on a minute! What I suggested is a mere theory. I don't know if I could actually carry out the required backwards steps even in the specific case of $\frac{47321}{33461}$.

I'm sure you could. But what you have just said is what's really important because it tells us what we should now work on before discussing anything else.

> Which is?

We should concentrate on the actual details of the backwards mechanism of working from the fraction $\frac{p}{q}$ back to its immediate predecessor.

A definite task. I know how to go forward via

$$\frac{p}{q} \rightarrow \frac{p+2q}{p+q}$$

but I would have to think hard about how to go backwards.

Well, knowledge of the forward process will show us how to go back if we set about it properly.

But how do we set about it properly?

Give the fraction preceding $\frac{p}{q}$ a temporary name, such as $\frac{r}{s}$; r for its numerator and s for its denominator. Then figure out how r, s, p, and q are related by using your understanding of how $\frac{r}{s}$ becomes $\frac{p}{q}$.

When you say what to do, it sounds as if there is nothing to it at all. Right; let me see if I can pull this off. Since the denominator of the new fraction is the sum of the old numerator and old denominator, it must be that

$$q = r + s$$

Very good.

And since the numerator of the new fraction is the sum of the old numerator and twice the old denominator, it must be that

$$p = r + 2s$$

So

$$p = r + 2s$$
$$q = r + s$$

which looks very like what my teacher used to call "simultaneous equations."

Exactly, but quite simple ones from which you'll figure out r and s in terms of p and q without too much trouble.

I don't know about that. I'm sure I have forgotten the tricks, so you'll have to help me.

Okay, so as to move matters along. If we subtract the second equation from the first, we get that

$$p - q = s$$

But this is the denominator s figured out already. That was painless.

Subtracting one equation from the other eliminated the r, as they say. Now it is an easy matter to get r in terms of p and q, since the first of your simultaneous equations says that

$$r = p - 2s$$

But there's a $2s$ in this equation.

Yes, which we'll now replace by $2(p - q)$ to get

$$r = p - 2(p - q) = p - 2p + 2q$$

or

$$r = 2q - p$$

and we're done.

Good. So

$$\frac{r}{s} = \frac{2q - p}{p - q}$$

is the fraction just before $\frac{p}{q}$.

True. Strictly speaking we have to say that $p - q$ cannot be zero.

Because division by 0 is forbidden.

Yes. As it happens $p - q = 0$ would mean $p = q$ and so $\frac{p}{q} = \frac{p}{p} = \frac{1}{1}$.

And we are not interested in looking for the fraction before this one.

Let's just check that $\frac{r}{s}$ becomes $\frac{p}{q}$ under the usual rule. Adding the numerator and denominator of the p and q version of the fraction $\frac{r}{s}$ gives

$$(2q - p) + (p - q) = q$$

as it should, since this is the denominator of $\frac{p}{q}$.

Let me add the numerator to twice the denominator of this same fraction to get

$$(2q - p) + 2(p - q) = 2q - p + 2p - 2q = p$$

which is the numerator of $\frac{p}{q}$ as it should be. This is great.

Removing the temporary scaffolding that was $\frac{r}{s}$, we may write the backwards mechanism as

$$\frac{2q - p}{p - q} \leftarrow \frac{p}{q}$$

Here we can think of the $\leftarrow$ as saying "goes back to," since this rule brings us backwards along the sequence.

Makes sense. I want to try out this "backwards rule" on $\frac{47321}{33461}$.

Well, then, off you go!

I should write this line left to right, shouldn't I? I get

$$\frac{19601}{13860} = \frac{2(33461)-47321}{47321-33461} \leftarrow \frac{47321}{33461}$$

which is not a fraction I recognize.

You'll just have to take more backwards steps. Soldier on until you hit one you recognize.

Okay, here we go again

$$\frac{577}{408} \leftarrow \frac{1393}{985} \leftarrow \frac{3363}{2378} \leftarrow \frac{8119}{5741} \leftarrow \frac{19601}{13860} \leftarrow \frac{47321}{33461}$$

At last we hit $\frac{577}{408}$, which I know is in the sequence.

Just as well this happened. If we keep applying the backwards rule we get

$$\frac{1}{1} \leftarrow \frac{3}{2} \leftarrow \frac{7}{5} \leftarrow \frac{17}{12} \leftarrow \frac{41}{29} \leftarrow \frac{99}{70} \leftarrow \frac{239}{169} \leftarrow \frac{577}{408}$$

which brings us back to the seed fraction $\frac{1}{1}$, as we'd expect.

What happens if we keep applying the backwards rule?

Try it and see.

I get

$$\cdots -\frac{7}{5} \leftarrow -\frac{3}{2} \leftarrow -\frac{1}{1} \leftarrow \frac{1}{0} \leftarrow \frac{1}{1}$$

So it looks as if the sequence reemerges after the $\frac{1}{0}$ fraction, but this time with minus signs before its terms. But didn't we say earlier that it's forbidden for a fraction to have a zero below the line because you are never allowed to divide by 0?

You are right, but let's just say that we needn't get distracted by what the backwards rule does with our sequence once it passes beyond the seed $\frac{1}{1}$.

All right, if you say so. I think by now I know how to backtrack along the sequence as well as go forward along it.

Which means that you are ready to return to the big question: Is every $\frac{p}{q}$ for which $p^2 - 2q^2 = \pm 1$ a member of the sequence?

Maybe, but first what do you mean when you write both a plus and a minus in front of the 1?

Well, it is pronounced "plus or minus one" and is shorthand for something that maybe either 1 or −1.

But not both at the same time?

Correct; it is either one or the other but never the two values at the same time.

Which I suppose is no more than you'd expect.

Exactly. The ±1 is just a very handy way of dealing with both possibilities in the same discussion. So now, maybe, we can use what we have learned to prove that the sequence is all-inclusive and that there are no maverick $\frac{p}{q}$'s.

I hope so, but you'll have to show me how.

We could do it by convincing ourselves that every fraction $\frac{p}{q}$ for which $p^2 - 2q^2 = \pm 1$ backtracks under the application of the backwards rule to the seed $\frac{1}{1}$.

But if we are not assuming $\frac{p}{q}$ is actually in the sequence, how can we be sure that the fraction obtained by the backwards rule has the ±1 property also?

A key point which we'll deal with right now. Why don't we calculate the quantity top squared minus twice bottom squared for this fraction and see what turns up?

For the fraction $\frac{2q-p}{p-q}$?

Yes. We get

$$
\begin{aligned}
(2q - p)^2 - 2(p - q)^2 &= 4q^2 - 4pq + p^2 - 2(p^2 - 2pq + q^2) \\
&= 4q^2 - 4pq + p^2 - 2p^2 + 4pq - 2q^2 \\
&= 2q^2 - p^2 \\
\Rightarrow (2q - p)^2 - 2(p - q)^2 &= -(p^2 - 2q^2)
\end{aligned}
$$

So $p^2 - 2q^2$ has popped out, but with a minus sign in front of it.

It has. This answer is most obliging because it tells us that if the quantity top squared minus twice bottom squared is 1 for $\frac{p}{q}$, then it is −1 for the preceding fraction obtained using the backwards rule.

And if the quantity top squared minus twice bottom squared is −1 for $\frac{p}{q}$, then it is the other way round for that same preceding fraction.

Yes, all because $(2q - p)^2 - 2(p - q)^2 = -(p^2 - 2q^2)$.

Simple, really. So whether or not $\frac{p}{q}$ is in the sequence, the fractions obtained from it by applying the backwards rule will all have the property that their numerator squared minus twice their denominator squared is ±1.

If $\frac{p}{q}$ has this property, yes, from what we have just shown. By the way, are you convinced that the backwards rule always produces a fraction at each stage and, if so, is it always a smaller fraction?

> Oops! I took all this for granted, partly, I suppose, because of the concrete examples. But surely both of these things are true. If p and q are integers, then so are $2q - p$ and $p - q$, and so one divided by the other is a fraction.

I think we'll allow this without any cross-examination.

> That's a relief! In the examples, $2q - p$ is always smaller than p, and $p - q$ is always smaller than q. Is it easy to show that this is always true for positive p and q?

It is, and perhaps you'll convince yourself of this privately.

> I promise. Now all we have to do is convince ourselves that the fraction $\frac{p}{q}$ leads back to the seed $\frac{1}{1}$ and not to something else.

How do you know that you can get back in a finite number of steps?

> How do you come up with these inconvenient little questions all the time? It's almost perverse.

I know. Some might say that it's an annoying habit acquired from hanging around too many super-careful mathematicians who examine every assumption.

> Worse than lawyers! But I'll see if I can answer it. The denominator q, no matter how large it is, is still a finite number. Now the backwards-process reduces this at every step, so it could, at the very worst, take q steps to go right back down to the bottom.

Argued like a mathematician. In fact, as we see from the examples, the descent is likely to be much more rapid than that.

> Well, I'm glad that's out of the way also. What I am wondering about is what else $\frac{p}{q}$ could eventually lead back to. I'm beginning to be certain of the result already. If the backwards-process ever taps into the sequence, then it must lead back to $\frac{1}{1}$.

Certainly, but how could you know this; or more importantly, what if it doesn't?

> If it doesn't, how could there be a parallel trail that leads back to a seed as small as $\frac{1}{1}$? If so, what could it be? I didn't see any other options when I was searching for near misses many moons ago.

Now I'm not sure I'm following your train of thought.

> The fraction $\frac{p}{q}$ has to track back to something that is an absolute minimum, like $\frac{1}{1}$, where a further application of the backwards rule breaks down.

In what sense breaks down?

> Well, where either the numerator or the denominator is no longer positive or maybe both are no longer positive. As we saw above, applying the backwards rule to $\frac{1}{1}$ gives the absurd $\frac{1}{0}$, where the denominator is no longer positive.

Please go on.

> We know that the backwards-process will lead from $\frac{p}{q}$ back down to a seed $\frac{a}{b}$ in a finite number of steps. What we have to do is prove that $\frac{a}{b}$ must be $\frac{1}{1}$.

If we can show this, then I'll be convinced.

> But how do we put what I have said into algebra to see if we can settle the question once and for all?

As we have discussed, the backwards-process will lead from $\frac{p}{q}$ back down to a seed $\frac{a}{b}$ in a finite number of steps, so let's examine what can we say about $\frac{a}{b}$

> That $a^2 - 2b^2 = \pm 1$.

That certainly, but also that a and b are both positive integers; but that either $2b - a \le 0$ and/or $a - b \le 0$, since $\frac{2b-a}{a-b}$ is the fraction preceding $\frac{a}{b}$ according to the backwards rule.

The inequality symbol $\le$ means less than or equal to.

> Because of what I said about $\frac{a}{b}$ being at the absolute bottom?

Yes, absolute bottom in terms of positive integers. If $2b - a$ and $a - b$ were also both positive integers, then since $\frac{2b-a}{a-b}$ is less than $\frac{a}{b}$, we'd have a contradiction.

> Because we'd have a positive fraction less than $\frac{a}{b}$, which is supposed to be the smallest such fraction.

Precisely. Now we just follow carefully where each of these separate inequalities leads.

> The inequalities $2b - a \le 0$ and $a - b \le 0$?

Yes. Let's take $2b - a \le 0$ first. This implies that $2b \le a$, and so, since a and b are both positive, we may say that $4b^2 \le a^2$.

> But can we not say this without having to say that a and b are both positive?

No. Great care must be taken when dealing with inequalities. For example, when the true inequality $2(-7) \le -13$ is squared on both sides, the inequality sign must be reversed to give $[2(-7)]^2 \ge (-13)^2$. If you fail to do this, you end up with the absurd $196 \le 169$.

The inequality symbol $\ge$ means greater than or equal to.

> So that's why you had to stress that a and b are both positive?

Yes. But when both quantities are positive, the inequality sign is preserved, as they say. Now

$$a^2 \geq 4b^2 \Rightarrow a^2 - 2b^2 \geq 4b^2 - 2b^2$$
$$\Rightarrow a^2 - 2b^2 \geq 2b^2$$
$$\Rightarrow \pm 1 \geq 2b^2$$

since we know that $a^2 - 2b^2 = \pm 1$.

I'll have to go slowly through this argument again for myself later so as to take in all the steps. But I'm happy to accept what it says so that we can get on with things.

Once you get the first line, the rest follow easily. Now $2b^2$ is at least 2, since b is a positive integer, so it's not possible for $2b - a \leq 0$.

So you are arguing by contradiction, like the Ancient Greeks.

Exactly! The numerator $2b - a$ does not go nonpositive in the backwards step from $\frac{a}{b}$ to $\frac{2b-a}{a-b}$.

So it's the other possibility $a - b \leq 0$ that must occur.

And that we now investigate.

$$a - b \leq 0 \Rightarrow a \leq b$$
$$\Rightarrow a^2 \leq b^2 \text{ (because } a \text{ and } b \text{ are both positive)}$$
$$\Rightarrow a^2 - 2b^2 \leq b^2 - 2b^2$$
$$\Rightarrow a^2 - 2b^2 \leq -b^2$$
$$\Rightarrow \pm 1 \leq -b^2$$

using $a^2 - 2b^2 = \pm 1$ again.

I can see that $1 \leq -b^2$ is impossible, because 1 is positive and $-b^2$ is strictly negative since b is a positive integer.

So what does that leave?

That $a^2 - 2b^2 = -1$ with $-1 \leq -b^2$. If I am not mistaken, $-1 \leq -b^2$ happens only if $b = 1$.

You are not mistaken. The only other possibility, $b = -1$, is ruled out because b is a positive integer.

But this is absolutely fantastic, because it says that b must be 1. I'm sure this means that a must be 1 also, and if so, $\frac{a}{b} = \frac{1}{1}$, which is what we want to prove.

And is $a = 1$?

Well, $b = 1$ and $a^2 - 2b^2 = -1$ give $a^2 - 2 = -1$ or $a^2 = 1$. This implies that $a = 1$, as a is a positive integer. We have it!

We have indeed. Marvelous!

It was no joke to prove that if $\frac{p}{q}$ is such that $p^2 - 2q^2 = \pm 1$, then it is a term of the sequence

$$\frac{1}{1}, \quad \frac{3}{2}, \quad \frac{7}{5}, \quad \frac{17}{12}, \quad \frac{41}{29}, \quad \frac{99}{70}, \ldots$$

It certainly wasn't, but we got there.

I'll have to ask easier questions from now on.

Segregation

We need to pause to catch our breath after the strenuous mental exertions of our latest enquiry.

Definitely. I for one wouldn't mind seeing the level drop down a bit.

Let's just savour for a little while what we now know about the sequence

$$\frac{1}{1}, \quad \frac{3}{2}, \quad \frac{7}{5}, \quad \frac{17}{12}, \quad \frac{41}{29}, \quad \frac{99}{70}, \quad \frac{239}{169}, \quad \frac{577}{408}, \ldots$$

first encountered on your searches for what you were later to call near misses.

Firstly, we know how to generate as many terms of the sequence as we might want by applying the rule

$$\frac{m}{n} \rightarrow \frac{m+2n}{m+n}$$

first to the seed $\frac{1}{1}$, then to the fraction it generates, and so on in turn to each new fraction.

For as long as we are prepared to continue.

Secondly, because of this simple rule, we understand why

$$m^2 - 2n^2 = \pm 1$$

for each fraction $\frac{m}{n}$ in the sequence.

Yes. We can explain how a certain characteristic of the seed fraction is passed on via the rule to each fraction in the sequence. Well, we should add, in a plus or minus fashion.

I really liked the argument that shows this to be the case.

And, as a result of our most recent, rather tortuous, escapade, we also know that the fractions in the sequence are the only ones satisfying $m^2 - 2n^2 = \pm 1$.

I thought that was a much tougher argument. I'll need to go over it again a number of times I'd say, before I'll be certain that I understand it fully.

We can put a slightly different slant on this result. As we have said before, the equation $m^2 - 2n^2 = \pm 1$ is equivalent to

$$m^2 = 2n^2 \pm 1$$

and says that the perfect square m^2 is within 1 of $2n^2$. This means that the numerators of the fractions in the sequence, and these alone, give the collection of perfect squares that are within 1 of twice another perfect square.

> And so give all the near misses, which is amazing.

It is indeed amazing how much your one observation about these near misses opened up.

> Maybe now we should figure out which of the fractions in the sequence make $m^2 - 2n^2 = -1$ and which make $m^2 - 2n^2 = 1$?

This is easily done. Every second fraction in the above sequence, beginning with the seed $\frac{1}{1}$, satisfies $m^2 - 2n^2 = -1$ because the seed fraction does and because this quantity alternates in sign as it moves from one fraction to the next in the sequence.

> Of course. This means the terms
>
> $$\frac{1}{1}, \frac{7}{5}, \frac{41}{29}, \frac{239}{169}, \dots$$
>
> of the main sequence make $m^2 - 2n^2 = -1$.

Yes. Because these terms obtained from the original, or main, sequence as you have just called it, form a sequence in their own right, this new sequence is often said to be a *subsequence* of the main sequence.

> Understood. On the other hand all the remaining terms make up another subsequence
>
> $$\frac{3}{2}, \frac{17}{12}, \frac{99}{70}, \frac{577}{408}, \dots$$
>
> in which $m^2 - 2n^2 = 1$.

So, we could say that -1 is the *signature* of each of the fractions in the first subsequence while 1 is the signature of all the terms in the second subsequence.

> Is there any point in making this distinction?

Yes. Consider a fraction $\frac{m}{n}$ from the primary sequence; if $m^2 - 2n^2$ works out to be -1, then we know it belongs to the first subsequence. Otherwise its signature works out to be 1, and it belongs to the second subsequence. It is a useful concept because, for example, it allows a computer program to test a fraction from the sequence to find out to which subsequence it belongs.

Very smart. I understand now. Would you give me a fraction so that I may test it?

One such fraction, not that far out in the main sequence, is $\frac{8119}{5741}$.

Okay, fraction, let me see what your signature is. The calculation

$$(8119)^2 - 2(5741)^2 = 65,918,161 - 2(32959081)$$
$$= 65,918,161 - 65,918,162$$
$$\Rightarrow (8119)^2 - 2(5741)^2 = -1$$

tells me you live somewhere out along the first subsequence because your signature is −1.

Correct. Can we say something significant about each of these subsequences?

If I remember correctly, we actually showed that the first three terms of the first subsequence provide better and better approximations to $\sqrt{2}$, but always underestimate it, and I was to show that the fourth fraction in this subsequence is an even better underestimate of $\sqrt{2}$, which it is. So we may say that

$$1 < \frac{7}{5} < \frac{41}{29} < \frac{239}{169} < \sqrt{2}$$

What might we conjecture the case in general to be?

That successive terms of the sequence

$$\frac{1}{1}, \frac{7}{5}, \frac{41}{29}, \frac{239}{169}, \dots$$

provide better and better underestimates of $\sqrt{2}$.

This seems plausible. Do you think you could prove it?

I'll take a stab at it. I'll begin by saying that each fraction $\frac{m}{n}$ in this sequence satisfies

$$m^2 = 2n^2 - 1$$

and I'll imitate the clever trick that you used before.

Which is?

To divide this last equation through by n^2 to get that

$$\left(\frac{m}{n}\right)^2 = 2 - \frac{1}{n^2}$$

You learn your lessons well, I see. This device shows the square of $\frac{m}{n}$ on the left-hand side. But don't let me interrupt you.

Straight away, we can say that each fraction in the sequence underestimates $\sqrt{2}$.

Yes, because when squared, each amounts to 2 minus the reciprocal of the positive quantity n^2.

Reciprocal?

The reciprocal of a quantity is just 1 divided by that quantity. I have interrupted you again, so let me say that I agree with all you have said so far.

Now the denominators of the fractions in this sequence increase, quite rapidly as you can judge from the first four fractions.

I must ask why you can be sure of this in general.

Because the denominator of a typical fraction in the primary sequence is the sum of the positive denominator and positive numerator of the previous fraction. The denominators get bigger and bigger as we move out along this sequence.

In such a way that they grow beyond all bounds. The phrase "tend to infinity" is often used to suggest this type of growth.

Has a ring to it!

Unfortunately, I must cut across your flow of thought for a moment to explain a subtlety that I refrained from mentioning earlier. Perhaps I should have.

Oh!

Looking at the rule

$$\frac{m}{n} \to \frac{m+2n}{m+n}$$

it is quite natural to assume that both numerators and denominators of the fractions in the primary sequence grow; but, unfortunately, there is a hidden assumption here.

"Hidden assumption" sounds serious.

How do we know that the fraction $\frac{m+2n}{m+n}$ doesn't reduce to a fraction whose denominator is smaller than the previous denominator? For example, suppose we changed this step, which follows the rule

$$\frac{7}{5} \to \frac{17}{12}$$

to this step, which does *not* follow the rule

$$\frac{7}{5} \rightarrow \frac{15}{12}$$

Then, since $\frac{15}{12} = \frac{5}{4}$, we'd end up with the next fraction having a smaller denominator than that of the fraction before it.

> I see the difficulty. In fact, I remember being surprised when the fraction $\frac{428571}{999999}$, with its long numerator and long denominator reduced to $\frac{3}{7}$ with just one digit both in its numerator and denominator. But what if we were to agree not to reduce the fractions at each stage; wouldn't we then be guaranteed that the denominators grow?

Yes, but then other things might go wrong. For example,

$$\frac{14}{10} = \frac{7}{5}$$

but the fraction $\frac{7}{5}$ is such that top squared minus twice bottom squared is $49 - 50 = -1$, whereas the fraction $\frac{14}{10}$ is such that top squared minus twice bottom squared is $196 - 200 = -4$.

> Another twist. Annoying and at the same time interesting. But am I right in saying that any time we generate a new fraction using the rule it is always in lowest form anyway?

You are right. And later we might show why the rule guarantees that this will always happen.

> So we now have another property that the rule passes from fraction to fraction?

Yes, when a fraction such as $\frac{1}{1}$ is the seed. So if we accept that this is so, your argument is a sound one and I'll let you get on with it.

> It seems that you can never be too careful in putting forward a mathematical argument. Anyway, to get back to what I was saying: as the denominators increase, the value of $\frac{1}{n^2}$ decreases, showing that the amount by which the squares of successive fractions of the subsequence underestimate 2 gets smaller.

Proving that these fractions provide better and better underestimates of $\sqrt{2}$. Excellent!

> Thank you. It seems to me that eventually the fractions of this sequence must come extremely close to $\sqrt{2}$.

Especially when the reciprocal of n^2 becomes very small. A phrase much used in mathematics is "arbitrarily close." It and words like "eventually" are quite difficult to make precise.

> But we get the general idea.

So we may write that

$$\frac{1}{1} < \frac{7}{5} < \frac{41}{29} < \frac{239}{169} < \cdots < \cdots \sqrt{2}$$

What about the other subsequence

$$\frac{3}{2}, \frac{17}{12}, \frac{99}{70}, \frac{577}{408}, \cdots$$

of the primary sequence?

In this case, each fraction $\frac{m}{n}$ makes

$$m^2 = 2n^2 + 1$$

So we can use your previous argument to show that successive fractions of this subsequence overestimate $\sqrt{2}$ by smaller and smaller amounts.

Yes, simply because of the +1 on the right-hand side of the equation instead of the previous −1.

Thus

$$\sqrt{2} \cdots < \cdots < \frac{577}{408} < \frac{99}{70} < \frac{17}{12} < \frac{3}{2}$$

If we combine this with the inequalities associated with the other subsequence, we may say that

$$\frac{1}{1} < \frac{7}{5} < \frac{41}{29} < \frac{239}{169} < \cdots < \cdots \sqrt{2} \cdots < \cdots < \frac{577}{408} < \frac{99}{70} < \frac{17}{12} < \frac{3}{2}$$

which is fairly impressive.

So we have shown that the subsequence

$$\frac{1}{1}, \frac{7}{5}, \frac{41}{29}, \frac{239}{169}, \cdots$$

of the main sequence

$$\frac{1}{1}, \frac{3}{2}, \frac{7}{5}, \frac{17}{12}, \frac{41}{29}, \frac{99}{70}, \frac{239}{169}, \frac{577}{408}, \cdots$$

provides a sequence of successive approximations to $\sqrt{2}$, which get closer and closer to $\sqrt{2}$ while always remaining less than it.

Yes. Why don't we term it the *under-subsequence* because it is a sequence of rational approximations to $\sqrt{2}$, each of whose terms underestimates $\sqrt{2}$.

So the subsequence

$$\frac{3}{2}, \frac{17}{12}, \frac{99}{70}, \frac{577}{408}, \cdots$$

is the over-subsequence because its provides a sequence of successive approximations to $\sqrt{2}$, which always overestimate it.

Yes, by getting closer and closer to $\sqrt{2}$ from above $\sqrt{2}$.

From above? Approaching $\sqrt{2}$ from the right-hand side on the number line?

That's what I mean. Now, since successive terms of the under-subsequence are getting closer and closer to $\sqrt{2}$ without ever exceeding it, it must be that this under-subsequence is always increasing, meaning that a term is always bigger than its predecessor.

All right, we have just proved this.

Although the under-subsequence is an increasing subsequence, the same is not true of the main sequence, or parent sequence, as we might also call it because we know that its terms continually see-saw from one side of $\sqrt{2}$ to the other.

Of course, because of the alternating property.

Now, if we imagine the numbers in the under-subsequence depicted by ultra-fine red dots on the number line, then reading from left to right, the first red dot is at 1, the next at 1.4, the next at 1.4137 . . . , and so on.

So you are imagining that the terms of the under-subsequence appear as red dots along the number line as we move out along the sequence.

Providing an infinite procession of red dots, which remain below the $\sqrt{2}$ point.

It's hard to imagine how they can all squeeze in between 1 and $\sqrt{2}$ if there is always a little gap between each of them.

A very good observation. The reason they do is because the gap between two successive red dots diminishes the further one goes out along the sequence.

A kind of proportional decreasing of gap size, as it were.

Not exactly, but this is something we could investigate another time. Let us just say that the gaps diminish very rapidly but without any ever becoming zero.

They'd simply have to, wouldn't they? Otherwise, how could all these fractions be different and be less than $\sqrt{2}$?

Quite so.

There must be quite a crowding of these red dots "just below" $\sqrt{2}$.

There is, and the closer to $\sqrt{2}$, the greater is the crowding, as you put it.

And, I suppose, if the numbers in the over-subsequence are represented by equally fine blue dots on the number line, beginning at the first point $\frac{3}{2} = 1.5$, then we observe a movement of blue dots from right to left as successive fractions of this decreasing sequence make their appearance.

Yes. With the blue dots growing closer and closer together as the terms of the sequence approach $\sqrt{2}$ from the right.

And no blue dot would ever pass to the left, beyond $\sqrt{2}$.

Never. They crowd closer and closer to the right of $\sqrt{2}$ without ever passing $\sqrt{2}$ or landing on it.

Intriguing behavior.

If you think about it, any finite interval of the line, no matter how large and how close to $\sqrt{2}$ but not containing $\sqrt{2}$, contains only a finite number of the terms from the sequence, while any interval containing $\sqrt{2}$ as an internal point, no matter how miniscule, must contain an infinity of terms from both of the subsequences.

I'm sure I would have to think about this statement for quite some time to take it all in, if I could at all. I'll settle for understanding that it's reds on the left and blues on the right.

And never the twain shall meet, prevented from ever mingling by the irrational barrier at $\sqrt{2}$. The number line harbors many mysteries.

It would seem so.

You might note that the increasing under-subsequence

$$\frac{1}{1}, \quad \frac{7}{5}, \quad \frac{41}{29}, \quad \frac{239}{169}, \ldots$$

is very different from the increasing sequence of denominators:

$$1, \quad 5, \quad 29, \quad 169, \ldots$$

In what way?

Both sequences increase, but the terms of the under-subsequence never exceed or even reach $\sqrt{2}$, while in the sequence of denominators one can find terms that exceed any specified finite limit.

Different types of increasing?

Yes, it serves as a warning that the phrase "getting larger and larger" does not necessarily mean increasing beyond all limits.

Which is what you'd be inclined to think in ordinary speech.

True. Our discussion of the under-subsequence shows that getting larger and larger may not mean this at all. However, a sequence whose terms tend to infinity must contain an infinite number of terms that exceed any finite number, no matter how big.

> Another statement I'd need time to think about. I thought you promised this session wouldn't be hard going.

Did I? By the way, we still have one small job to do.

> Which is?

We must explain why a fraction generated by the rule never has to be reduced when the seed is $\frac{1}{1}$.

> I had forgotten.

No Reductions

When the rule

$$\frac{m}{n} \to \frac{m+2n}{m+n}$$

is applied to the seed fraction $\frac{1}{1}$, and in turn to each new fraction generated, the sequence

$$\frac{1}{1}, \quad \frac{3}{2}, \quad \frac{7}{5}, \quad \frac{17}{12}, \quad \frac{41}{29}, \quad \frac{99}{70}, \quad \frac{239}{169}, \quad \frac{577}{408}, \ldots$$

is obtained.

> As we well know by now.

When generated in this way, each fraction is always in reduced form.

> It never happens then that the numerator and denominator have a factor in common that can be canceled?

No reductions are ever needed. The fraction is always born in lowest form.

> How can you be sure of this?

Well, that's precisely what I must convince you of. However, I should say that this is true of each fraction depends vitally on the same property being possessed by the seed $\frac{1}{1}$.

> You mean that it itself is in lowest form?

Yes, the numerator and denominator sharing no factor other than the trivial factor 1.

> So with a different seed, it could happen that the fractions generated by the rule have numerators and denominators that have factors in common?

Factors other than 1, yes—nontrivial factors as they are called.

Do you have an example?

Let us try the rule on the seed $\frac{4}{2}$, which you'll notice is not in reduced form.

I do. The numerator and denominator share the non-trivial factor 2.

That's right. I am deliberately choosing not to reduce the fraction to its lowest form. Applying the rule gives

$$\frac{4+2(2)}{4+2} = \frac{8}{6}$$

as the next fraction.

I notice that the numerator and denominator of this fraction have exactly the same nontrivial common factor, 2, as the seed fraction $\frac{4}{2}$.

Yes—the one and only. Reapply the rule on this nonreduced fraction to generate the next fraction in the sequence.

We get

$$\frac{8+2(6)}{8+6} = \frac{20}{14}$$

What is the highest common factor of the numerator and denominator in this case?

The same as before, 2.

The first few terms of the sequence generated are

$$\frac{4}{2}, \frac{8}{6}, \frac{20}{14}, \frac{48}{34}, \frac{116}{82}, \frac{280}{198}, \frac{676}{478}, \ldots$$

You might like to verify that, for each fraction, the number 2 is the highest common factor of the numerator and denominator.

I can see already that this is the case.

If we reduce the seed $\frac{4}{2}$ to its lowest terms, $\frac{2}{1}$, before applying the rule successively to each fraction generated, then we get

$$\frac{2}{1}, \frac{4}{3}, \frac{10}{7}, \frac{24}{17}, \frac{58}{41}, \frac{140}{99}, \frac{388}{239}, \ldots$$

as the first few terms generated without ever having to reduce them.

Each is born in lowest form, as you said a moment ago.

Precisely. If we start the generation process with the seed $\frac{30}{18}$, we get

$$\frac{30}{18}, \frac{66}{48}, \frac{162}{114}, \frac{390}{276}, \frac{942}{666}, \frac{2274}{1608}, \frac{5490}{3882}, \cdots$$

What do you say about this sequence?

> The highest common factor of the numerator and denominator in the seed $\frac{30}{18}$ is 6, so I'd be inclined to think that the same is true for each of the other fractions.

Which means that we should find that each of the numerators is divisible by 6, with the same being true of the denominators.

> It is. Canceling the 6 common to each numerator and denominator gives the sequence

$$\frac{5}{3}, \frac{11}{8}, \frac{27}{19}, \frac{65}{46}, \frac{157}{111}, \frac{379}{268}, \frac{915}{647}, \cdots$$

> It looks to me as if all these fractions are in lowest form.

They are. So have you gleaned enough from these examples to make a conjecture regarding seeds and the sequences they generate via constant application of the propagation rule?

> I think so. It seems to me that the highest common factor of the numerator and denominator of each fraction generated must be exactly the same as highest common factor of the seed's numerator and denominator.

Which, if true, explains why the fractions in our original sequence with seed $\frac{1}{1}$ are always born in lowest form.

> It would, because the numerator 1 and denominator 1 of the seed $\frac{1}{1}$ have only the factor 1 in common. But how are you going to prove that this theory is true?

By returning to the rule

$$\frac{m}{n} \rightarrow \frac{m+2n}{m+n}$$

and showing that the highest common factor, or greatest common divisor as it is also called, of m and n in the fraction $\frac{m}{n}$ is exactly the same as the greatest common divisor of $m + 2n$ and $m + n$ in the fraction $\frac{m+2n}{m+n}$.

> I can see that if you could do this, it would explain everything. But how you are going to do it?

It takes a little thinking and some experience with the divisibility properties of whole numbers.

> Divisibility properties? Sounds highbrow to me.

The idea is simple. Show that the numbers m and n have the same common factors as the numbers $m + 2n$ and $m + n$.

> Maybe I could do it with numbers, but definitely not with letters.

All right, let us try it with numbers to get the general idea. We saw a minute ago that

$$\frac{m}{n} = \frac{30}{18} \Rightarrow \frac{m+2n}{m+n} = \frac{30+2(18)}{30+18} = \frac{66}{48}$$

The fact that 2 divides $m = 30$ and $n = 18$ makes it a certainty that 2 divides $m + n = 30 + 18$.

> You mean there is no need to add them and check that 2 divides the answer exactly?

Precisely. Because 2 divides the individual parts 30 and 18, it divides their sum. Agreed?

> Seems right.

The fact that 2 divides $m = 30$ and $n = 18$ also makes it a certainty that 2 divides $m + 2n = 30 + 2(18)$.

> Because 2 divides the individual parts 30 and 2(18), it has to divide their sum $30 + 2(18)$.

Yes. So here 2 being a common factor of $m = 30$ and $n = 18$ ensures that it is also a common factor of $m + 2n$ and $m + n$.

> I think I'm happy with this.

Now the same is also true for the other common factor, 3, of $m = 30$ and $n = 18$ for exactly the same reasons, would you not agree?

> I'm sure I will, when I take in all of what you are saying.

In this case then, we have shown that any common factor of m and n is automatically a common factor of $m + 2n$ and $m + n$.

> I'll accept this. But what now?

The next bit is somewhat harder to see. We are going to show that any common factor of $m + 2n = 66$ and $m + n = 48$ is also a common factor of $m = 30$ and $n = 18$.

> But you are not going to show this directly, simply by checking.

No. What I'm going to do may strike you as a little odd, but here goes. First

$$m = 2(m + n) - (m + 2n)$$

as you can easily check.

$2(m + n) - (m + 2n)$
$= 2m + 2n - m - 2n$
$= m$

> You mean in general or just for $m + n = 48$ and $m + 2n = 66$?

In general. It is certainly true for these numbers—look:

$$30 = 2(48) - 66 = 96 - 66$$

I see. So?

So we can say that the common factor 2 of 48 and 66 divides 30 because it divides both 2(48) and 66 and so must divide the difference $2(48) - 66$, which is 30.

A very odd way to show that 2 divides 30.

Certainly in the case of a specific number such as 30, but not when we come to the general argument involving $m + 2n$ and $m + n$.

Because you don't have specific numbers to work on.

Yes. For exactly the same reasons, the common factor 3 of $m + 2n = 66$ and $m + n = 48$ is a factor of $m = 30$. So what we have shown at this stage is that the common factors 2 and 3 of $m + 2n = 66$ and $m + n = 48$ are also factors of $m = 30$.

All right.

Now we show that the common factors 2 and 3 of $m + 2n = 66$ and $m + n = 48$ are also factors of $n = 18$.

And, as already said, you don't do this simply by verifying that each divides into $n = 18$ exactly.

No, we must show this using the fact that the numbers $m + 2n = 66$ and $m + n = 48$ are divisible by these numbers. Can you see how to do it?

I don't think so. I'll leave it to you.

Well, it is easy to check that

$$n = (m + 2n) - (m + n)$$

in general. In particular, $18 = n = (m + 2n) - (m + n) = 66 - 48$.

I can see that both are true.

Now, since 2 and 3 divide each of 66 and 48, they divide their difference 18. Thus, in this case, any common factor of $m + 2n$ and $m + n$ is also a factor of n.

A difference is just like a sum.

For divisibility purposes, yes. We have shown that the common factors of $m + 2n = 66$ and $m + n = 48$ are also common factors of $m = 30$ and $n = 18$.

And?

We already showed the other way round, that the common factors of m and n are also common factors of $m + 2n$ and

$m + n$. This means, for these specific numbers at any rate, that m and n have exactly the same set of common factors as $m + 2n$ and $m + n$.

All of which means?

That they must have the same highest common factor.

The crunch point. Is the general argument more complicated?

No. It is virtually the same as the one given. In one direction you show that any common factor of m and n must also be a common factor of $m + n$ and $m + 2n$. This is almost obvious.

Because m and n are in both sums?

Yes. Now, in the other direction, you use the equations

$$m = 2(m + n) - (m + n)$$
$$n = (m + 2n) - (m + n)$$

to argue that any common factor of $m + 2n$ and $m + n$ must be a factor of both m and n.

And so is a common factor of m and n.

This means that the pair of numbers m and n share exactly the same common factors as the pair $m + 2n$ and $m + n$.

Therefore, their greatest common divisor must be the same.

So we have another case of the rule

$$\frac{m}{n} \rightarrow \frac{m+2n}{m+n}$$

passing a property of the seed to successive generations.

The Two-Steps Rule

May I ask a question concerning the under- and over-subsequences?

By all means. What is it?

What are the rules relating to each of these sequences?

You mean rules that tell us how to go from the typical fraction to its successor?

Yes.

Actually, the same rule applies to both.

The same rule for the increasing sequence and the decreasing sequence?

The very same.

Even though the terms of the under-subsequence, the red fractions, all live to the left of $\sqrt{2}$, whereas all the terms of over-subsequence, the blue fractions, live to the right of $\sqrt{2}$.

Yes again.

I'm intrigued!

It is not so surprising when you recall that both sequences are subsequences of the main one

$$\frac{1}{1},\ \frac{3}{2},\ \frac{7}{5},\ \frac{17}{12},\ \frac{41}{29},\ \frac{99}{70},\ \frac{239}{169},\ \frac{577}{408},\ \ldots$$

consisting as they do of its alternate terms.

But what then gives the two subsequences such different characteristics?

Their different starting values, or seeds.

So the two sequences spring from different seeds but obey the same rule of generation. Is that it?

In a nutshell. To begin, I suggest that we focus on finding the rule for the under-subsequence.

But you said the same rule holds for the over-subsequence.

I did and do, but that this must be so is not clear to you at this stage. So let us start without any preconceived notions.

All right.

How are the terms from the under-subsequence obtained from the parent sequence?

By selecting from it the first term, the third, the fifth, and so on.

So if we could discover a rule that brings us from a typical term in the parent sequence to the term two beyond it, then we'd know the rule for selecting the terms of the under-subsequence, provided we started at the first term.

I think I may see now why the same rule will select the over-subsequence from the parent sequence.

Why?

By starting at the second fraction instead of the first, the same selection procedure will pick out all the even terms. The new rule is like learning how to take two steps of a ladder at a time and doesn't depend on where you start out on the ladder.

You could say that and call it "the two-steps rule."

I suppose we'll need to use our original rule, which tells us how to take one step on the ladder of the original sequence.

Certainly, the "one-step" rule

$$\frac{m}{n} \rightarrow \frac{m+2n}{m+n}$$

which, you will remember, tells how the fraction following $\frac{m}{n}$ is obtained from it in terms of m and n. Now ask yourself, what is the fraction after this one?

> This is a new departure—to have to consider another term—a third term.

It is, but three terms are involved in this case because we are skipping over an in-between term to get to the next one.

> I can see the logic in that.

With the main sequence of fractions we need to focus only on two terms because knowledge of the current term is enough to find the next fraction.

> Right, so must I now calculate the term after $\frac{m+2n}{m+n}$ in terms of m and n also?

If you would.

> So the fraction $\frac{m+2n}{m+n}$ is my starting fraction for forming the next one?

Exactly. And what do you get?

> Well, the new denominator is the sum of the current numerator and denominator, and so is
>
> $$(m + 2n) + (m + n) = 2m + 3n$$

New bottom is old bottom plus old top, isn't that it?

Indeed.

> The new numerator is the current numerator plus twice the current denominator, and so is
>
> $$(m + 2n) + 2(m + n) = 3m + 4n$$

New top is old top plus twice old bottom.

Right again. So you now have the top and the bottom of the new fraction.

> Which means that the next fraction is
>
> $$\frac{3m+4n}{2m+3n}$$
>
> if all the above calculations are correct.

They're right on.

> But what now?

What now? You have obtained the rule for turning one fraction of the subsequences into the next one.

I have?

Undoubtedly. Ask yourself what it is that you have just shown.

I must collect my thoughts. That if $\frac{m}{n}$ is a typical fraction in the main sequence, then

$$\frac{m+2n}{m+n} \quad \text{and} \quad \frac{3m+4n}{2m+3n}$$

are the next two fractions following it in that order.

Absolutely correct. So what's the rule of formation for either or both of the subsequences?

Is it

$$\frac{m}{n} \to \frac{3m+4n}{2m+3n}$$

because to get the next term in either of the subsequences we must skip over a term in the main sequence?

You have it. This is the rule that does the trick for both subsequences. Why don't you check it out on the under-subsequence

$$\frac{1}{1}, \ \frac{7}{5}, \ \frac{41}{29}, \ \frac{239}{169}, \ldots$$

by testing it on the seed $\frac{1}{1}$?

I can't wait. Setting $m = 1$ and $n = 1$ in the general rule gives

$$\frac{1}{1} \to \frac{3+4}{2+3} = \frac{7}{5}$$

which is the second term of this sequence. It works!

Now see if the fraction $\frac{7}{5}$ acting as $\frac{m}{n}$ in the newly discovered rule gives the third term of the sequence.

Setting $m = 7$ and $n = 5$ gives

$$\frac{7}{5} \to \frac{3(7)+4(5)}{2(7)+3(5)} = \frac{41}{29}$$

which is the third term of the under-subsequence. Pretty impressive.

You should now check that

$$\frac{m}{n} \to \frac{3m+4n}{2m+3n}$$

gives the second term of the over-subsequence

$$\frac{3}{2}, \frac{17}{12}, \frac{99}{70}, \frac{577}{408}, \ldots$$

when it is applied to its seed $\frac{3}{2}$.

Of course I should. Setting $m = 3$ and $n = 2$ gives

$$\frac{3}{2} \rightarrow \frac{3(3)+4(2)}{2(3)+3(2)} = \frac{17}{12}$$

as I was fairly sure would happen, but it's nice to see it pop out all the same.

Repeat the procedure to see if you get the next term.

I don't doubt it. Setting $m = 17$ and $n = 12$ gives

$$\frac{17}{12} \rightarrow \frac{3(17)+4(12)}{2(17)+3(12)} = \frac{99}{70}$$

which is term number three of the over-subsequence.

Now that we have the general rule for generating both the under- and over-subsequences, I want to pose a puzzle whose answer you must give me without doing any calculations.

I'm not sure I like the sound of this, but let it not be said that I ducked a challenge.

If we calculate the quantity

$$(\text{top})^2 - 2(\text{bottom})^2$$

for the fraction

$$\frac{3m+4n}{2m+3n}$$

what will the answer be?

So you want me to tell you what

$$(3m + 4n)^2 - 2(2m + 3n)^2$$

simplifies to *without* doing a single calculation?

Yes.

So there must be a quick trick or observation that answers this. If I do give the right answer, I'll want you to work out the above expression by hand afterward.

Fair enough; that will be your reward, but you must give me the right answer.

I know. So $(\text{top})^2 - 2(\text{bottom})^2$ is the square of the numerator minus twice the square of the denominator. Ah yes! For the main sequence we proved that this quantity is always either -1 or 1. Why are you humming to yourself?

Am I? I thought I was listening in silence to a great mind thinking aloud.

Muddled mind is more like it. So let me see, where am I? This quantity is -1 for every term in the under-subsequence and 1 for every term in the over-subsequence.

What nice weather we are having!

Let's hope my thoughts are as clear as the day is. In algebraic terms, $(\text{top})^2 - 2(\text{bottom})^2$ is $m^2 - 2n^2$ for the typical fraction $\frac{m}{n}$ in the main sequence. Now $m^2 - 2n^2 = -1$ for all the odd terms in the main sequence, and $m^2 - 2n^2 = 1$ for all the even terms. So what is all this telling me I'd get if I were to work out $(3m + 4n)^2 - 2(2m + 3n)^2$?

Perhaps I should take a little stroll to leave you to your thoughts.

No need to go. I think I know the answer.

I'm dying to hear it.

Does it work out at $m^2 - 2n^2$?

It does, but why?

Well, you have heard my thoughts up to now.

Yes, with great satisfaction.

Thank you. If I work out $(3m + 4n)^2 - 2(2m + 3n)^2$ in either subsequence, I must get the same as I would for the previous fraction because $(\text{top})^2 - 2(\text{bottom})^2$ never changes for either of these subsequences.

And?

But $(\text{top})^2 - 2(\text{bottom})^2$ is $m^2 - 2n^2$ for the previous fraction $\frac{m}{n}$.

Top class!

Now you must verify it by doing all the algebra.

Gladly. Well, $(3m + 4n)^2 = 9m^2 + 24mn + 16n^2$. Agreed?

I must think about the middle term for a second. It's just twice $3m \times 4n = 12mn$?

Correct. While $(2m + 3n)^2 = 4m^2 + 12mn + 9n^2$ implies that $2(2m + 3n)^2 = 8m^2 + 24mn + 18n^2$.

I'm still with you.

Therefore

$$(3m + 4n)^2 - 2(2m + 3n)^2 = (9n^2 + 24mn + 16n^2)$$
$$- (8m^2 + 24mn + 18n^2)$$
$$= m^2 - 2n^2$$

showing that a little thinking can often save a lot of hard working out.

It is still nice to check, just in case the argument is flawed.

Quite right. It is all too easy to overlook something.

So even though the tops and bottoms of the fractions change from term to term in the under-subsequence and the over-subsequence, the quantity

$$(\text{top})^2 - 2(\text{bottom})^2$$

remains constant all the time.

Amid a sea of changing numbers, as it were. This quantity is an example of an *invariant*.

Something that never varies?

Yes, perpetuates itself without change along the sequence.

So, for the under-subsequence, the value of this invariant is -1, and for the over-subsequence, it is 1.

As we said a little while back, the invariant -1 is the signature of the under-subsequence and the invariant 1 the signature of the over-subsequence.

The main sequence doesn't possess $m^2 - 2n^2$ as an invariant since this quantity does not remain constant but alternates between -1 and 1 as we move along the terms.

Yes, strictly speaking.

The Pell Sequence

Our recent discussion concerning the two-steps rule governing the under- and over-subsequences of the sequence

$$\frac{1}{1}, \frac{3}{2}, \frac{7}{5}, \frac{17}{12}, \frac{41}{29}, \frac{99}{70}, \frac{239}{169}, \frac{577}{408}, \ldots$$

helped me solve a puzzle you gave me some time ago.

It did? That's great. Please give me all the details.

You remember the drill sergeant parading squadrons?

Of course. We pretended that you were this eccentric person with a desire for perfect square formations.

The one and only, with a squadron consisting of an ideal
number of soldiers, a perfect square number.

Who, as soon as the worthy band had been trained to march
faultlessly in a perfect square formation, was informed by
the top brass that the squadron was to be doubled. Thereby
denying the sergeant, unwittingly no doubt, the possibility of
parading the larger squadron in the same fashion.

Yes. The most that could be hoped for was near-perfection. That
twice the original number of soldiers would either be one more
or less than a perfect square.

If I remember rightly, you discovered that if the original
squadron size is the square of any of the numbers in the
sequence of denominators

$$1, \quad 2, \quad 5, \quad 12, \quad 29, \quad 70, \quad 169, \quad 408 \ldots$$

formed from the main sequence of fractions just listed, then
the enlarged squadron is just 1 off a perfect square.

In fact, the squares of the corresponding numbers in the
sequence of numerators

$$1, \quad 3, \quad 7, \quad 17, \quad 41, \quad 99, \quad 239, \quad 577 \ldots$$

of the main sequence of fractions give the size of the squadron
doubled to within plus or minus 1.

And of course we now know that these are the only numbers
that work. All in all, a wonderful discovery!

The puzzle you set me was in relation to the sequence of
denominators

$$1, \quad 2, \quad 5, \quad 12, \quad 29, \quad 70, \quad 169, \quad 408 \ldots$$

which you said is known as the Pell sequence.

I remember. Your task was to find a rule, presuming there is
one, that allows us to calculate successive terms of this
sequence from previous ones without having to refer to any
other sequence, such as the main sequence of fractions. A very
nice exploration, if I may say so, this time with whole numbers
as opposed to fractions.

Yes. Initially I felt things would be simpler than up to now, since
I thought whole numbers would have to be simpler to deal with
than fractions.

With both their tops and their bottoms to be considered.
So how did you do?

Well, I was wrong in thinking that the task would be simpler
just because it involved whole numbers.

So that was a lesson in itself.

Eventually, I got the rule for the Pell sequence and the numerator sequence

$$1, \quad 3, \quad 7, \quad 17, \quad 41, \quad 99, \quad 239, \quad 577, \ldots$$

and I want you to check if my explanations hold up.

I'm all ears!

Starting with the Pell sequence

$$1, \quad 2, \quad 5, \quad 12, \quad 29, \quad 70, \quad 169, \quad 408, \ldots$$

I noticed, after many false starts, that

$$70 = 2 \times 29 + 12$$
$$29 = 2 \times 12 + 5$$
$$12 = 2 \times 5 + 2$$
$$5 = 2 \times 2 + 1$$

This suggests that the next term is twice the previous one plus the term before that.

Very well observed: not an observation that everybody would see without a lot of searching.

You've got that right! It also took me a long time to see because originally I was looking for a connection between just two terms, a term and the one before it.

And?

After many unsuccessful and frustrating attempts I sensed that I was on the wrong track. I couldn't find anything until I started to look at a term and the *two* terms before it.

So you abandoned a lost cause and took a bold new leap.

I don't know about that. Let's just say that I was primed in some sort of way to do what I did because of our recent investigation, which involved dealing with three terms rather than two.

You had an idea that was in the air, as it were. So tell me more.

But this rule I have just mentioned doesn't apply to the first two terms.

Well, how could it?

I know, but for a while I was worried about that. Then it dawned on me that I was being unreasonable because there aren't two previous terms in these two cases.

It's as simple as that.

> Talking about ideas being in the air, it occurred to me, again I suppose because of our previous experience with the under- and over-subsequences, to check if the same rule held for the sequence of numerators as well.

Very enterprising. Because if you don't mind my saying so, the two cases might not strike some as being that similar.

> I know. It was only afterward that I asked myself whether I was justified in doing what I did.

After all, the numerator and denominator sequences are constructed from *every* term in the main sequence, whereas the under- and over-subsequences are constructed from alternate terms. So the fact that one rule held for the under- and over-subsequences doesn't necessarily imply that the numerator and denominator sequences will also be governed by a single rule.

> Anyway, at the time, I just tried my "Pell rule," as I call it, on the other sequence without really thinking anything too much about what I was doing.

And *does* the Pell rule apply to the numerator sequence

$$1, \quad 3, \quad 7, \quad 17, \quad 41, \quad 99, \quad 239, \quad 577, \ldots$$

also?

> Like a dream! Of course, I know I should allow myself to say no more than, "I think so" until you check my reasoning. Here's the evidence provided by the four terms after the first and second of this sequence

$$239 = 2 \times 99 + 41$$
$$99 = 2 \times 41 + 17$$
$$41 = 2 \times 17 + 7$$
$$7 = 2 \times 13 + 1$$

> This much was enough to make me believe that the Pell rule holds in general for this sequence as well.

It *looks* at this stage as if you've struck gold.

> Because you have been showing me how to argue algebraically, I wanted to explain why this rule holds in general for generating the terms of both sequences beyond the first two.

Very ambitious, but admirable.

> I think part of my problem with trying to show things in general by means of algebra is that I don't know where to start.

You are not unique in that regard. It can be hard even for seasoned campaigners, because the explanations for some observations can be very far from the surface.

I think I have discovered that, luckily, this is not the case here.

I have every confidence in you.

I thought to myself, "The observation was easy to make, so surely the explanation will be easy to find."

Would that this were always the case! Mathematics is laced with observations that children can make but whose proofs are still awaited.

Are you serious?

Totally. One of the greatest mathematicians of all time tells us that he often discovered results empirically that took him months to prove. He also said that, if he so wished, he could write down countless conjectures that people could neither prove nor disprove.

That's intriguing, but I won't ask you for examples right now because I don't want to get distracted from what we are doing.

Which is to examine your explanation of the Pell rule.

Right. I said to myself that we must work with what we know. Perhaps, because the numerator and Pell sequences both spring from the main sequence

$$\frac{1}{1}, \ \frac{3}{2}, \ \frac{7}{5}, \ \frac{17}{12}, \ \frac{41}{29}, \ \frac{99}{70}, \ \ldots$$

I gathered what I know in general about this sequence to see if I could pick up any leads.

A much relied-on strategy.

To begin, the one-step rule for generating successive terms of the primary sequence beginning with the seed $\frac{1}{1}$ is

$$\frac{m}{n} \rightarrow \frac{m+2n}{m+n}$$

where, as we said many times before, $\frac{m}{n}$ stands for a typical term in the sequence.

Correct. Fire away.

The fraction

$$\frac{m+2n}{m+n}$$

is the typical one that follows $\frac{m}{n}$. I then asked myself if anything else that we know about this sequence might be useful.

And?

The fact that $m^2 - 2n^2$ being always either -1 or 1 came to mind, but I couldn't see how I could use it on what I was doing.

Then?

I thought of our most recent result, the two-steps rule

$$\frac{m}{n} \rightarrow \frac{3m+4n}{2m+3n}$$

This gives the fraction immediately after the fraction $\frac{m+2n}{m+n}$ in the primary sequence.

Being two fractions further forward in this sequence than $\frac{m}{n}$.
How was this important in your quest?

The minute I thought of it, I knew I was on the right track because it puts the three terms

$$\frac{m}{n}, \quad \frac{m+2n}{m+n}, \quad \frac{3m+4n}{2m+3n}$$

into the picture.

And this is significant?

It is exactly what I needed. My supposed rule relates a term to the two before it. I knew I had reached a critical point, and it only remained to be seen if I could think straight enough to find the explanation I sought.

So how did you pilot your ship into the harbor?

I focused on the denominators. I wanted to show that any denominator is obtained by adding twice the previous denominator to the one just before that.

Provided you exclude the first two terms, isn't that it?

Yes. So let me extract the denominators from the three successive fractions

$$\frac{m}{n}, \quad \frac{m+2n}{m+n}, \quad \frac{3m+4n}{2m+3n}$$

and take it from there.

Let me assist you. They are

$$n, \quad m+n, \quad 2m+3n$$

taken in increasing order.

Now what my Pell rule says about the third or final term, $2m + 3n$, is that it is equal to twice the previous term $m + n$ added to *its* previous term, which is n.

And is it?

The simple calculation

$$2(m + n) + n = 2m + 2n + n = 2m + 3n$$

shows that it is.

Magnificent! I can find no fault in this argument. Your explanation of why the Pell rule holds in general looks sound to me.

I thought so, but I wanted to be sure, to have you check through it with me just in case.

How about the numerator sequence?

The same rule holds, since

$$2(m + 2n) + m = 2m + 4n + m = 3m + 4n$$

shows that the third numerator is twice the second numerator plus the first numerator.

It is interesting that the two sequences have exactly the same structure. We might say that the numerator sequence is a cousin of the Pell sequence.

The Pell sequence and its cousin require a *pair* of seeds each to get them growing.

They are generated from a different pair by exactly the same growth mechanism. Their apparent difference is only superficial and is due to their different "initial values," as these seeds can be called.

Different seed, but same breed.

Well, I think we can safely say that you established the truth of your Pell rule in fairly short order once you spotted it from an examination of the numerical evidence available to you. After that, it was only a matter of translating what you suspected to be the case into symbols to provide a convincing argument as to why the rule holds in general.

As soon as I sensed that I was on the right track, I felt sure I would be able to explain everything.

Congratulations. A great achievement for one whose algebra was a little rusty.

You're being kind; nonexistent might be a better description.

Whatever; you're getting the hang of it. It takes time and thought to arrive at that level where you're able to go from start to finish, and this you've done in fine style.

I have to admit that it is a real thrill to be able to show why something is true in general using algebra, particularly when you simply guess that this is the case after examining a small amount of numerical evidence.

The power of algebra—great for converting insight into hindsight.

CHAPTER 4

Witchcraft

I want to show you something that might strike you as witch-craft. To begin, let's go right back to the equation

$$\sqrt{2} \times \sqrt{2} = 2$$

which tells us . . .

. . . in a simple way exactly what is meant by $\sqrt{2}$.
Watch how I use this relationship to multiply $\sqrt{2} - 1$ by $\sqrt{2} + 1$:

$$
\begin{array}{rcr}
\sqrt{2} & - & 1 \\
\sqrt{2} & + & 1 \\
\hline
2 & - & \sqrt{2} \\
& + \sqrt{2} & - 1 \\
\hline
2 & + \quad 0 & - 1
\end{array}
$$

Can you tell me where it was used in this multiplication?

At the very first step, when you write down 2 beneath the first line as the result of multiplying $\sqrt{2}$ by $\sqrt{2}$.

Exactly. Alternatively, we might perform the above calculation this way:

$$
\begin{aligned}
(\sqrt{2} - 1)(\sqrt{2} + 1) &= [\sqrt{2} \times (\sqrt{2} + 1)] - [1 \times (\sqrt{2} + 1)] \\
&= [(\sqrt{2} \times \sqrt{2}) + (\sqrt{2} \times 1)] - [(1 \times \sqrt{2}) + (1 \times 1)] \\
&= [2 + \sqrt{2}] - [\sqrt{2} + 1] \\
&= 1
\end{aligned}
$$

Either way we get the same result.

Which is just as well.

Both of these calculations tell us that

$$(\sqrt{2} - 1)(\sqrt{2} + 1) = 1$$

Now watch out for the magic I'm about to demonstrate using this relationship.

I can't wait.

Dividing across by $\sqrt{2}+1$ gives

$$\sqrt{2}-1=\frac{1}{\sqrt{2}+1}$$

which, in turn, gives

$$\sqrt{2}=1+\frac{1}{1+\sqrt{2}}$$

This is the point of departure for the sorcery to come.

A strange-looking equation, if you don't mind my saying so.

Actually, it is called an *identity*, since both sides are identical as numbers. It is a little out of the ordinary, certainly, because it says something about $\sqrt{2}$ in terms of itself.

Since $\sqrt{2}$ is on both sides of the equation?

Yes. I'm now about to infuse a little imagination into the proceedings by applying some mathematical sleight of hand to this identity—the wizardry I promised.

Great.

I'm going to replace the $\sqrt{2}$ on the right-hand side of the identity by 1.

Just on the right-hand side, not on the left?

On the right-hand side only, which means that the expression will no longer be an identity or an equation.

So what does it become?

An expression where the right-hand side provides what I hope is a reasonable approximation to $\sqrt{2}$.

I'll have to pay attention to see how this works.

Indeed, since 1 is a fairly poor approximation to $\sqrt{2}$ by any standards, we may not end up with anything spectacular, but let's see.

Let me do the calculation. When I replace the $\sqrt{2}$ on the right-hand side of

$$\sqrt{2}=1+\frac{1}{1+\sqrt{2}}$$

with a 1, this side gives the fraction

$$1+\frac{1}{1+1}=\frac{3}{2}$$

which looks familiar.

It should. It is the second fraction in our sequence

$$\frac{1}{1},\ \frac{3}{2},\ \frac{7}{5},\ \frac{17}{12},\ \frac{41}{29},\ \frac{99}{70},\cdots$$

Of course!

Now we know already that this fraction is a better approxima-tion to $\sqrt{2}$ than the 1 with which we began.

So you used the identity to improve on the approximation 1 of $\sqrt{2}$? Impressive.

You could express it this way. We can say that we began with the approximation

$$\sqrt{2}\approx1$$

and improved it to

$$\sqrt{2}\approx\frac{3}{2}$$

$\approx$ means "is approximately"

Does this give you any ideas?

How about using this new estimate for the $\sqrt{2}$ on the right-hand side of the identity

$$\sqrt{2}=1+\frac{1}{1+\sqrt{2}}$$

as we did with 1 a moment ago to see what comes out?

Just what I was hoping you would say.

Let me do it. The right-hand side becomes

$$1+\frac{1}{1+\frac{3}{2}}=1+\frac{1}{\frac{5}{2}}=1+\frac{2}{5}=\frac{7}{5}$$

The next fraction in the sequence!

And an improved approximation of $\sqrt{2}$. All of which is very interesting, would you not agree?

So much so that I must play the same trick again. We get

$$1+\frac{1}{1+\frac{7}{5}}=1+\frac{1}{\frac{12}{5}}=1+\frac{5}{12}=\frac{17}{12}$$

which is the fourth term in the sequence.

And so a new improved approximation to $\sqrt{2}$.

> It looks as if we are generating our original sequence in a different way.

It does. The very first crude approximation of 1 can be thought of as the fraction $\frac{1}{1}$, the first fraction in the sequence. Do you want to convince me that this process generates our sequence?

> Do I have a choice? Give me a hint as to how to start.

Imagine that the fraction $\frac{p}{q}$ is the one most recently generated by the procedure, and take it from there.

> Okay. So instead of continuing with the $\frac{17}{12}$ just obtained, I imagine that we have generated as far as the fraction $\frac{p}{q}$, a typical term in the sequence being generated using the "strange" identity?

Yes. Again, we don't use $\frac{m}{n}$ so as to avoid making any assumptions.

> Doing exactly as above with $\frac{p}{q}$ instead of a specific approximating fraction to $\sqrt{2}$, the next term is given by

$$1+\cfrac{1}{1+\cfrac{p}{q}} = 1+\cfrac{1}{\cfrac{q+p}{q}}$$

$$= 1+\cfrac{q}{p+q}$$

$$= \cfrac{(p+q)+q}{p+q}$$

$$= \cfrac{p+2q}{p+q}$$

> which is exactly the same rule as before.

No difference other than p where we previously had m, and q instead of n.

> Since the seed is also $1 = \frac{1}{1}$, the sequence generated by this new procedure is the same as our original sequence.

We have discovered the same sequence of approximations to $\sqrt{2}$ in another way.

What If?

> But what happens if we choose a different starting approximation to $\sqrt{2}$ on the right-hand side of

$$\sqrt{2} = 1+\cfrac{1}{1+\sqrt{2}}$$

And say we used even a completely off-the-wall approximation.

You can see from what you have just done that the generating rule is still the same, so in essence we'd just be choosing a different seed. The real question then is: will the successive terms of the sequence generated from this seed by applying the rule over and over still approach $\sqrt{2}$?

Okay.

Let's experiment a little and see.

Right. I'll go crazy and take a seed of 10.

A much-admired and often-used approximation of $\sqrt{2}$!

No doubt! Let me get to work. Beginning with $p = 10$ and $q = 1$ gives

$$\frac{p+2q}{p+q} = \frac{10+2}{10+1} = \frac{12}{11}$$

so the next term is $\frac{12}{11}$.

Do you notice anything about the size of this new term?

It is very close to 1 which, I suppose, probably means that the next term will be close to our previous $\frac{3}{2}$.

Why do you say this?

Because putting $\frac{1}{1}$ into the rule gives $\frac{3}{2}$, so, since $\frac{12}{11}$ is close to 1, I assume that when it is put into the rule something close to $\frac{3}{2}$ should come out.

I see your point. Let's have a look then.

Updating p to 12 and q to 11 gives

$$\frac{p+2q}{p+q} = \frac{12+2(11)}{12+11} = \frac{34}{23}$$

as the third term of the new sequence.

It may not look it, but this fraction *is* close to $\frac{3}{2}$, as you suspected.

I'm going to calculate some more terms in this new sequence. The next term has denominator $34 + 23 = 57$, while its numerator is $34 + 2(23) = 80$.

Which means that the term after this has denominator $80 + 57 = 137$, while its numerator is $80 + 2(57) = 194$.

Adding these terms has this new sequence starting out with

$$\frac{10}{1}, \frac{12}{11}, \frac{34}{23}, \frac{80}{57}, \frac{194}{137}, \ldots$$

Well, it certainly looks as if this sequence, with the exception of the first "absurd" entry, is very close on a term-by-term basis to the sequence

$$\frac{1}{1}, \frac{3}{2}, \frac{7}{5}, \frac{17}{12}, \frac{41}{29}, \frac{99}{70}, \ldots$$

Because of this I'd be surprised if this new sequence does not also provide successive approximations that get closer and closer to $\sqrt{2}$.

I'm inclined to agree with you.

I'm going to sin and get the decimal equivalents to five places. Here they are.

10.00000,   1.09090,   1.47826,   1.40350,   1.41605 . . .

Definitely heading for $\sqrt{2}$.

How can you be sure?

I'll bet my life on it, and that you can prove I'm right.

Such confidence! Why don't we do as we did before, which was to square some of the fractional approximations and see how close they are to 2.

To avoid using decimals. For discovering possible proofs, you like to stick with whole numbers wherever possible?

Yes. Easier to pick up a scent, as it were.

Which number should I square first?

Well $10^2 = 100$ is so far from 2 as to be laughable, so why not begin with the next term.

Which is $\frac{12}{11}$. Now

$$\begin{aligned}
\left(\frac{12}{11}\right)^2 &= \frac{144}{121} \\
&= \frac{242 - 98}{121} \\
&= 2 - \frac{98}{121}
\end{aligned}$$

I see you have used that old trick again in your calculation, which reveals that the square of the second term in the new sequence is 2 minus the fraction $\frac{98}{121}$.

But surely this couldn't be considered a small error, so the second term isn't worth much as an approximation of $\sqrt{2}$.

No, but you wouldn't expect it to be.

Let me tackle the third term, $\frac{34}{23}$. I get

$$\left(\frac{34}{23}\right)^2 = \frac{1156}{529}$$
$$= \frac{1058+98}{529}$$
$$= 2 + \frac{98}{529}$$

That 98 has popped up again. Why do I suspect that this is not a coincidence?

You tell me; but first, do we have an improved approximation in $\frac{34}{23}$?

Yes, because the error $\frac{98}{529}$ is smaller than the previous error of $\frac{98}{121}$.

Agreed. This time the fractional error measures an excess, as opposed to a shortfall specified by the previous one. This approximation to $\sqrt{2}$ hardly sets the world on fire either.

I know, it's still way off. I'll test the next term $\frac{80}{57}$ and keep an eye out for the appearance of the mysterious 98.

Do.

Right; here goes:

$$\left(\frac{80}{57}\right)^2 = \frac{6400}{3249}$$
$$= \frac{6498-98}{3249}$$
$$= 2 - \frac{98}{3249}$$

There's that 98 again, this time with a minus sign.

So $\frac{80}{57}$ is an improvement on all previous approximations because $\frac{98}{3249}$ is the smallest fractional error obtained so far.

It is, but it's no great shakes, either.

No indeed.

But as we continue out along the sequence, the terms should improve in the same way they do in our original sequence.

Which is something we are in the process of proving, if I'm not mistaken.

Of course. I had better start by making some general observations based on what we have just been doing.

Or more accurately conjectures, which may point the way forward.

For starters, it seems to me that as approximations, the terms bounce around as they do for the sequence that begins with $\frac{1}{1}$.

Can you be more precise?

Successive approximations jump between being overestimates of $\sqrt{2}$ to being underestimates, all the while improving.

As you say, the same alternating pattern as before.

But not quite the same. This time the first term is an absurd *over*estimate of $\sqrt{2}$, whereas the more moderate 1 seeding the other sequence underestimates $\sqrt{2}$.

Point taken. And what about this mysterious 98, as you termed it?

I think I can explain why it keeps turning up.

Show me.

If I'm right, it has to do with the quantity $p^2 - 2q^2$, the value of $(\text{top})^2 - 2(\text{bottom})^2$ for the general term $\frac{p}{q}$.

I'd say you're on the right track.

We showed already that if the next term in the sequence is

$$\frac{p+2q}{p+q}$$

then the quantity $(\text{top})^2 - 2(\text{bottom})^2$ has the value $2q^2 - p^2 = -(p^2 - 2q^2)$ for this term.

We did, and so?

Well, we know that if $\frac{p}{q}$ is a typical term in the new sequence, then the fraction $\frac{p+2q}{p+q}$ is the next term.

Yes.

And isn't it precisely this rule which guarantees that the value of the quantity $(\text{top})^2 - 2(\text{bottom})^2$ simply changes sign as one goes from term to term?

The very one. But where does the 98 come from?

From that crazy first term 10, or $\frac{10}{1}$. When $\frac{p}{q} = \frac{10}{1}$, the quantity $p^2 - 2q^2 = 100 - 2(1) = 98$.

Very good. And because $p^2 - 2q^2$ is always either this value or minus it, the number 98 propagates all along the sequence, appearing now as 98 and next as −98 and so on. So what now?

Because of what we have just been saying, I think I can prove that *any* sequence formed using the above rule must have its successive terms get closer and closer to $\sqrt{2}$.

You can, for any sequence no matter how absurd the initial value?

I think so. Let me tell you my thoughts.

You are making great strides.

To begin, no matter how badly chosen the first fraction is, it fixes the value of $p^2 - 2q^2$ forever for that sequence.

Well, to within a plus or a minus sign. "Up to sign" is how it is expressed, meaning that it is always either some particular number or its opposite.

If a is the value of $p^2 - 2q^2$ when this quantity is positive, then if $p^2 - 2q^2 = a$ for the initial choice of $\frac{p}{q}$, this quantity will alternate between a and $-a$ as $\frac{p}{q}$ moves along the sequence in question.

And the other way round if $p^2 - 2q^2 = -a$ for the initial choice of $\frac{p}{q}$.

Now can't we write that

$$p^2 - 2q^2 = \pm a$$

as we did for the case $a = 1$?

We can.

Now divide this equation through by q^2, as you did before, to get that

$$\left(\frac{p}{q}\right)^2 = 2 \pm \frac{a}{q^2}$$

This relationship shows that the square of the fraction $\frac{p}{q}$ is equal to 2, give or take $\frac{a}{q^2}$.

And what are you going to make of this?

Doesn't it mean that as q gets larger and larger, the quantity $\frac{a}{q^2}$ gets smaller and smaller?

It does, no matter what the value of a, as long as you are sure that q tends to infinity.

But doesn't it for the very same reason as before, which is that the next denominator is the sum of the previous numerator and denominator?

And since these numerators and denominators are positive integers, the denominators increase beyond all bounds.

So as q grows larger and larger, the quantity $\frac{a}{q^2}$ becomes smaller and smaller.

No matter what value *a* has?

> Yes. Even if *a* were as large as 10 million, say, the *q* values will eventually grow beyond this value. Then q^2 is much larger still and so makes $\frac{a}{q^2}$ into a tiny fraction.

Thus, no matter how large *a* is, the quantity $\frac{a}{q^2}$ eventually becomes so small that it can be considered negligible?

> If what I'm saying is correct. When the denominators *q* are very large, the corresponding fractions of the sequence have squares that are very close to 2.

Showing that successive terms of the sequence of fractions approach $\sqrt{2}$.

> By my reasoning.

And no matter what the intial $\frac{p}{q}$?

> Yes, provided it's a positive fraction, I suppose.

So any sequence generated by the rule

$$\frac{p}{q} \rightarrow \frac{p+2q}{p+q}$$

consists of terms that successively approach $\sqrt{2}$, irrespective of the starting term?

> I think so.

Something you have argued must be so, and most skillfully it must be said.

> Thank you. I really enjoyed that, but I'd like to investigate a little further to examine a hunch I have.

So, another exploration?

Always Between 1 and 2

> How about starting with an even more absurd initial approximation, just to see how the first few terms come out.

So what ridiculous value are you going to choose?

> Why not 1000?

Another well-known approximation of $\sqrt{2}$!

> A joke approximation I know, but my theory is that we'll still get quite good approximations to $\sqrt{2}$ after no more than a few fractions, using the usual rule.

Nothing to do but see immediately if what you think will happen does happen.

With $p = 1000$ and $q = 1$, the rule gives

$$\frac{p+2q}{p+q} = \frac{1000+2}{1000+1} = \frac{1002}{1001}$$

as the second term in the sequence whose seed is 1000.

What do you make of this?

It fits in with my hunch. This new value is close to 1, just like the value $\frac{12}{11}$ we got with the less crazy starting value of 10.

In fact, it looks very close to 1.

So from now on things shouldn't be that much different from the two previous sequences. Because this second term is down around 1, the successive terms of this sequence should make their way toward $\sqrt{2}$ at about the same rate as the corresponding terms in the previous two sequences.

Whether they do or not, and I believe that they will as you say, you have already shown that successive terms of the sequence must approach $\sqrt{2}$ eventually.

I'm going to go the opposite way now and choose an absurdly small approximation of $\sqrt{2}$, say the fraction $\frac{1}{1000}$. With $p = 1$ and $q = 1000$, the rule gives

$$\frac{p+2q}{p+q} = \frac{1+2000}{1+1000} = \frac{2001}{1001}$$

as the second term in the sequence that begins with $\frac{1}{1000}$.

This time you get a fraction which is just a little bit below 2.

This is just fine also, because another application of the rule will get us into the "settling down" stage, if I may call it that.

The next fraction is

$$\frac{2001+2(1001)}{2001+1001} = \frac{4003}{3002}$$

which is about $\frac{4}{3}$.

Between the $\frac{7}{5}$ and $\frac{17}{12}$ of the original sequence. So this sequence, with its very poor seed of $\frac{1}{1000}$, is up and running.

So do these numerical experiments bear out your hunch?

I think so. My hunch is this:

The second term of any sequence formed using the rule

$$\frac{p}{q} \rightarrow \frac{p+2q}{p+q}$$

is always a number between 1 and 2, no matter how its seed is chosen.

I think we'll promote this to the status of a conjecture. If this educated guess is true, then it goes some way to understanding why the successive terms of all the sequences so formed approach $\sqrt{2}$.

As usual, I'm not sure where to start the algebra to try to prove it. You'll have to help me out, once again.

Just enough to get you started. You have used the phrase "no matter how its seed is chosen."

I did.

By which you mean any conceivable rational number seed?

When I say *any* seed I suppose I mean any one of all the possible fractions.

So give this general rational seed a name.

Ah, right. May I call it $\frac{a}{b}$?

Anything except $\frac{p}{q}$, really. You cannot use $\frac{p}{q}$ because that algebraic expression already has the job of denoting the typical fraction of the sequence.

Could be confusing. So $\frac{a}{b}$ it is. But didn't we already use $\frac{a}{b}$ to stand for a seed?

We did. Appropriate, considering that a and b are the initial letters of the alphabet.

In this case, then

$$\frac{a+2b}{a+b}$$

plays the role of the general second term.

Precisely. Now you're set up.

Maybe, but what do I do now?

Express what it is you would like to prove in terms of $\frac{a}{b}$.

Oh, I see; a good idea. I'm saying that no matter how $\frac{a}{b}$ is chosen, the next term

$$\frac{a+2b}{a+b}$$

is always a term between 1 and 2.

That's it. Now you are getting a handle on it.

But I still don't know what to do.

Undoubtedly we are at the hardest stage, where we need to make some connection between what we are asserting and what we know.

This connection had better jump up and hit me.

It will if we can see the right way of looking at your conjecture: "Is always a number between 1 and 2," you say.

Yes, but what of it?

Can't we say then that it is a number of the form 1 plus some number that is less than 1. One plus a proper fraction. A proper fraction is one where the numerator is smaller than the denominator.

What is an improper fraction?

I remember. However, I'm still not being hit by any flash of insight.

Think back. Quite recently we came upon an expression that is of the form 1 plus something.

That's right. It was

$$1+\cfrac{1}{1+\cfrac{p}{q}}$$

—a relationship we came across when using the identity.

Exactly.

And we showed then, using a little algebra, that

$$1+\cfrac{1}{1+\cfrac{p}{q}}=\frac{p+2q}{p+q}$$

Yes, and which for the purposes of this discussion is none other than the term immediately after the seed, if we replace p by a and q by b.

How convenient. There's surely something here if only I could see it.

I'm sure you will. Try to use the fact that the term after the seed $\frac{a}{b}$ can also be written as

$$1+\cfrac{1}{1+\cfrac{a}{b}}$$

Since this is 1 plus something, all I have to do is convince the world that this something, namely,

$$\frac{1}{1+\dfrac{a}{b}}$$

is less than 1.

You had better prove that this quantity is greater than 0 also, otherwise it might end up subtracting from 1 instead of adding to it.

Oops! I automatically assumed that $\frac{a}{b}$ is positive.

Quite natural. It doesn't have to be, but the truth of your conjecture probably depends on its being so.

All possible positive seeds is what I had in mind. When this is the case, then $1 + \frac{a}{b}$ is greater than 1.

Agreed, because something positive is being added to 1.

So

$$\frac{1}{1+\dfrac{a}{b}}$$

is also a positive number.

Granted—the reciprocal of a positive number is also a positive number.

In this number, the numerator is less than the denominator, so it is a positive fraction that is less than 1.

Provided $\frac{a}{b}$ is positive.

Yes. I realize the importance of this assumption now. So

$$\frac{a+2b}{a+b}=1+\frac{1}{1+\dfrac{a}{b}}$$

is 1 plus a positive fraction less than 1. This proves my hunch.

Spell out exactly why.

Because 1 plus a positive fraction less than 1 is a fraction between 1 and 2.

I can't argue with that.

And this fraction $\frac{a+2b}{a+b}$ is the term that comes immediately after the general seed $\frac{a}{b}$. So the second approximation that the rule generates is always a number between 1 and 2, no matter where the positive seed comes from.

You have surpassed yourself.

Not at all; you led me along by the nose.

Perhaps I put you on the right trail because I felt its starting point isn't at all obvious.

Well, I'd never have found it, but I am delighted with where it has taken us. For me this discussion has been another example of how well algebra can explain things in general.

That's very pleasing to hear. What you have so ably demonstrated is that whether we begin with a very big estimate, a very small estimate, or even a moderate estimate $\frac{a}{b}$ of $\sqrt{2}$, our procedure reaches, after at most two steps, an estimate between 1 and 2. When this stage is reached, the rule begins to produce approximations of $\sqrt{2}$ that get better and better quite quickly.

I still find it hard to believe that we can start with any estimate whatsoever of $\sqrt{2}$, no matter how far off the mark it is, to kickstart the approximation procedure, and that it will zoom in on a very good rational approximation to $\sqrt{2}$ after six or seven steps.

The old adage, "A good start is half the battle" doesn't really apply here, because no matter how wildly we start the procedure, it simply rights itself on the next step.

And then it's business as usual.

I must mention that we could have used a slightly more direct argument to show that the fraction $\frac{a+2b}{a+b}$, which comes after the positive seed $\frac{a}{b}$, must always be a number between 1 and 2. I deliberately had you use the observation

$$\frac{a+2b}{a+b} = 1 + \frac{1}{1+\dfrac{a}{b}}$$

because it had arisen earlier and so allowed you to get on with your argument without having to perform any fresh algebraic manipulations.

But?

Well, many would regard the above starting point as slightly eccentric—although it does the job. It might be considered more normal to write

$$\frac{a+2b}{a+b} \overset{!}{=} \frac{(a+b)+b}{a+b}$$

$$= \frac{a+b}{a+b} + \frac{b}{a+b}$$

$$\Rightarrow \frac{a+2b}{a+b} = 1 + \frac{b}{a+b}$$

But I wouldn't have known how to do these manipulations.

Which is why I used something we had encountered before.

> I see you have put an exclamation mark over the first equal sign to indicate that it is a clever step.

It is the vital starting point for the next two steps, which get us to an equation that can also be used to prove your assertion. Can you see how?

> Time to put on my thinking cap again. If a and b are both positive, which I know they are, then the numerator b in the fraction $\frac{b}{a+b}$ is less than its denominator $a + b$. This means that the fraction itself is a positive fraction less than 1.

Exactly. And so 1 plus this fraction must give a fraction between 1 and 2.

> I understand. I suppose this argument is a little shorter than the one we gave.

Before we finish this particular discussion which began with your taking outlandish initial approximations for $\sqrt{2}$, I want you to try another bold experiment.

> Which is?

Your initial approximations were fractions, which was only right and proper, because our purpose is to find rational approximations to $\sqrt{2}$—particularly good ones. But what would happen if we applied the rule to the seed $\sqrt{2}$ itself?

> But how could the fraction $\frac{a}{b}$ be equal to $\sqrt{2}$?

It can't in the normal meaning of the word "fraction" as we have been using it since, as we well know, $\sqrt{2}$ is not a rational number. But what happens if we apply the rule with $a = \sqrt{2}$ and $b = 1$.

> Which is just saying that

$$\sqrt{2} = \frac{\sqrt{2}}{1}$$

Yes, a device we have used before.

> I'll try it to see. We get

$$\sqrt{2} = \frac{\sqrt{2}}{1} \to \frac{\sqrt{2}+2(1)}{\sqrt{2}+1}$$

So under the rule

$$\frac{\sqrt{2}}{1} \to \frac{2+\sqrt{2}}{\sqrt{2}+1}$$

Yes. Do you think you could manipulate the expression on the right-hand side to see if it might reduce to something simpler?

Where do I start? There doesn't seem much to go on.

Use the definition of $\sqrt{2}$.

Okay. Using the fact that $\sqrt{2} \times \sqrt{2} = 2$, we can replace the 2 in the numerator by $\sqrt{2} \times \sqrt{2}$ to get

$$\frac{2+\sqrt{2}}{\sqrt{2}+1} = \frac{\sqrt{2} \times \sqrt{2} + \sqrt{2}}{\sqrt{2}+1} = \frac{\sqrt{2}(\sqrt{2}+1)}{\sqrt{2}+1} = \sqrt{2}$$

So under the rule

$$\frac{\sqrt{2}}{1} \rightarrow \sqrt{2}$$

or more simply

$$\sqrt{2} \rightarrow \sqrt{2}$$

Amazing, under the rule the number becomes itself again!

Yes. Can you give a reason why this happens?

When the rule is applied to a fraction that approximates $\sqrt{2}$, it produces a new approximation to $\sqrt{2}$, which is closer to $\sqrt{2}$ and so is a better approximation. But if we start out with the exact value of $\sqrt{2}$, then the rule cannot improve on this, so it just sends the number into itself.

Sounds plausible. So the sequence generated by the rule when the seed is $\sqrt{2}$ is as follows:

$$\sqrt{2}, \quad \sqrt{2}, \quad \sqrt{2}, \quad \sqrt{2}, \quad \sqrt{2}, \quad \sqrt{2} \ldots$$

It's a "constant" sequence because every term is the same.

So you could say that there is no movement in this case?

Or say that the rule leaves the number $\sqrt{2}$ fixed.

Does it leave any other numbers fixed?

Yes. One other. You might like to practice your algebra by trying to find it.

Some other time, maybe. If I were to apply the rule in the same way to another irrational number, such as $\sqrt{3}$, as we did to $\sqrt{2}$, what would happen?

Or even to a number like π. Let me just say that if you were prepared to apply the rule about six or seven times, you'd end up getting expressions involving $\sqrt{3}$ or π that would be fairly good approximations to $\sqrt{2}$.

An exploration for another time, perhaps.

A Bold Leap of Imagination

I want to return to the identity

$$\sqrt{2} = 1 + \frac{1}{1+\sqrt{2}}$$

which we have already mined to produce the sequence

$$\frac{1}{1}, \frac{3}{2}, \frac{7}{5}, \frac{17}{12}, \frac{41}{29}, \frac{99}{70}, \frac{239}{169}, \frac{577}{408}, \ldots$$

in a different way from how you originally discovered it.

> Where we replaced the $\sqrt{2}$ on the right-hand side by 1 and so on.

Now I'm going to use the identity as a starting point again, to walk a different path, as it were.

> Let's bring it on, then.

Our first step is a bold imaginative one. Watch! This time I'm going to replace the $\sqrt{2}$ on the right-hand side of the identity not by an approximation, as we did previously, but by something exact.

> And this something exact is?

The entire right-hand side of the identity, which its left-hand side tells us is also $\sqrt{2}$.

> I'm not sure I follow what you're saying.

I am going to replace the $\sqrt{2}$ that lives on the right-hand side of the identity by

$$1 + \frac{1}{1+\sqrt{2}}$$

which is also equal to $\sqrt{2}$, is it not?

> Let me think. It is, because of the $\sqrt{2}$ on the left-hand side of the identity.

So I may replace $\sqrt{2}$ on the right-hand side of the identity by the term just displayed.

> Ah! I now see what you are about. Bold and imaginative indeed.

We get

$$\sqrt{2} = 1 + \frac{1}{1 + \left(1 + \frac{1}{1+\sqrt{2}}\right)}$$

which might strike you as a little eccentric.

> To say the least!

When the expression on the right-hand side is simplified we get

$$\sqrt{2} = 1 + \cfrac{1}{2 + \cfrac{1}{1 + \sqrt{2}}}$$

which is a little tidier.

But really strange.

Maybe, but now we get more daring. We replace the $\sqrt{2}$ nestled at the bottom right of this three-tier expression in the same way as we did a moment ago. After all, there is nothing to prevent us doing so.

Insane!

An even more perverse thought is to replace this humble $\sqrt{2}$ by the entire expression appearing on the right-hand side of the new identity just displayed, but that would be to follow a different path.

A madman's outing for another day?

Perhaps. Sticking with our first plan, we get the expression

$$\sqrt{2} = 1 + \cfrac{1}{2 + \cfrac{1}{1 + \left(1 + \cfrac{1}{1 + \sqrt{2}}\right)}}$$

which simplifies to

$$\sqrt{2} = 1 + \cfrac{1}{2 + \cfrac{1}{2 + \cfrac{1}{1 + \sqrt{2}}}}$$

This is an even more curious specimen than the previous one.

With four tiers you could say. I've never seen an expression such as this before. The right-hand side looks like a slanted ladder of fractions inclined at about 45 degrees.

But if you examine it for a while, you'll see that it has a simple structure. If we repeat the step just taken, what will we get?

Pretty much the same thing, but with an extra tier or rung of the form

$$2 + \frac{1}{}$$

coming before the final

$$2+\cfrac{1}{1+\sqrt{2}}$$

at the bottom.

Yes, the overall expression is

$$\sqrt{2}=1+\cfrac{1}{2+\cfrac{1}{2+\cfrac{1}{2+\cfrac{1}{1+\sqrt{2}}}}}$$

Here we have a longer ladder, as you refer to it, with three 2s down its left-hand side, whereas the previous ladder had two.

I suppose there is nothing to stop us *continuing* this substitution business *ad infinitum*?

Theoretically, nothing at all. In fact, if we imagine that this has been done, we get what is called the infinite *continued fraction expansion* of $\sqrt{2}$**:**

The ∵ mean that the pattern continues indefinitely.

$$\sqrt{2}=1+\cfrac{1}{2+\cfrac{1}{2+\cfrac{1}{2+\cfrac{1}{2+\cdots}}}}$$

This entire expression has rather an attractive elegance to it.

Apart from the first 1 immediately after the equal sign, it is all 2s down along one side and 1s down the other to infinity.

Yes. Bringing that first 1 to the other side of the equation gives

$$\sqrt{2}-1=\cfrac{1}{2+\cfrac{1}{2+\cfrac{1}{2+\cfrac{1}{2+\cdots}}}}$$

a form that is, perhaps, more pleasing in appearance.

Maybe, but as I said before, both expressions are unlike anything I have ever seen.

Aside from its infinite extent, this continued fraction expansion of $\sqrt{2}$, and that of $\sqrt{2}-1$, has a very simple structure. The same pattern appears over and over again through each layer of the descending expansion. So besides its almost bewilderingly simple definition, $\sqrt{2}$ has another form of simplicity, which it reveals in its continued fraction expansion.

> Some might not think the continued fraction expansion quite that simple, but it is very interesting, certainly.

Almost exotic. I think this infinite continued fraction expansion is credited to the Italian mathematician Raphael Bombelli, who wrote it down around 1572.

> And it wasn't known before?

I cannot say. Certainly the mathematical symbol $\sqrt{}$ for square roots was not long in use at the time.

> Once you are shown how the continued fraction is obtained, it becomes hard to believe that it was not always around.

All of which should encourage us to explore for ourselves whenever the mood takes us.

> You mean we might still find a nugget?

Maybe, you never know; but more for the pure thrill of discovering for oneself.

> A nugget in itself.

Another Manifestation

Now I'd like to show you something else that might offer you a challenge.

> Oh dear!

It will be quite straightforward. We saw that if we replace the $\sqrt{2}$ on the right-hand side of

$$\sqrt{2}=1+\frac{1}{1+\sqrt{2}}$$

with 1, then this side changes to

$$1+\frac{1}{1+1}=1+\frac{1}{2}$$

which is the fraction $\frac{3}{2}$.

> I remember.

But this expression is also the infinite continued fraction expansion of $\sqrt{2}$:

$$1+\cfrac{1}{2+\cfrac{1}{2+\cfrac{1}{2+\cfrac{1}{2+\cdots}}}}$$

truncated before the second plus sign or, if you prefer, after the first 2. With everything else thrown away, as it were.

Let me examine this. Okay, I see it.

Now if we replace the $\sqrt{2}$ on the right of the three-tier expression

$$\sqrt{2} = 1 + \cfrac{1}{2+\cfrac{1}{1+\sqrt{2}}}$$

with 1, what do we get?

Why don't I compute to find out? Substituting 1 for $\sqrt{2}$ on the right-hand side gives

$$1+\cfrac{1}{2+\cfrac{1}{1+1}} = 1+\cfrac{1}{2+\cfrac{1}{2}}$$

$$= 1+\cfrac{1}{\cfrac{5}{2}} = 1+\cfrac{2}{5}$$

$$= \frac{7}{5}$$

which—surprise, surprise—is the next fraction in the sequence

$$\frac{1}{1}, \frac{3}{2}, \frac{7}{5}, \frac{17}{12}, \frac{41}{29}, \frac{99}{70}, \frac{239}{169}, \frac{577}{408}, \ldots$$

after $\frac{3}{2}$.

The infinite continued fraction expansion of $\sqrt{2}$ truncated before the third plus sign is the third fraction in the sequence.

So

$$1+\frac{1}{2} = \frac{3}{2} \quad \text{and} \quad 1+\cfrac{1}{2+\cfrac{1}{2}} = \frac{7}{5}$$

shows these fractions in a different light. Is the challenge then to prove that the pattern continues?

Exactly, whichever way we choose to look at it. The fourth fraction in the sequence is obtained by replacing the $\sqrt{2}$ on the right-hand side of the four-tier expression

$$\sqrt{2} = 1 + \cfrac{1}{2 + \cfrac{1}{2 + \cfrac{1}{1 + \sqrt{2}}}}$$

with 1, or simply by truncating the infinite continued fraction expansion of $\sqrt{2}$ before the fourth plus sign to get

$$1 + \cfrac{1}{2 + \cfrac{1}{2 + \cfrac{1}{2}}}$$

since both expressions are equivalent. Check that you get the fraction $\frac{17}{12}$.

I see how to do this in an efficient manner. Write the *first* 2 in the expression as $1 + 1$ to get

$$1 + \cfrac{1}{1 + \left(1 + \cfrac{1}{2 + \cfrac{1}{2}} \right)}$$

I have placed big brackets around what I know from our previous work represents the fraction $\frac{7}{5}$.

Clever!

Substituting the $\frac{7}{5}$ gives

$$1 + \cfrac{1}{1 + \cfrac{7}{5}} = \frac{17}{12}$$

as the value of the four-tier expression or the infinite continued fraction truncated before the fourth plus sign.

Very accomplished!

This last equation also shows that the four-tier ladder number representing the fractional approximation $\frac{17}{12}$ is the same as the one obtained by substituting the third approximation $\frac{7}{5}$ for $\sqrt{2}$ on the right-hand side of two-tier starting identity

$$\sqrt{2} = 1 + \cfrac{1}{1 + \sqrt{2}}$$

which surprises neither of us.

If you follow the argument you have just given step by step, what we want to prove should become clear. In fact, we can say that, to date, we have shown that

$$\frac{1}{1} = 1$$

$$\frac{3}{2} = 1 + \frac{1}{2}$$

$$\frac{7}{5} = 1 + \cfrac{1}{2 + \cfrac{1}{2}}$$

$$\frac{17}{12} = 1 + \cfrac{1}{2 + \cfrac{1}{2 + \cfrac{1}{2}}}$$

because the first entry is obtained from the infinite continued fraction expansion of $\sqrt{2}$ on truncating it before the first plus sign. Doing so gives the first term of the sequence and so lends consistency to the whole conjecture.

I suppose we must now prove algebraically that successive truncations of the infinite continued fraction are just different ways of expressing the successive fractions of our original sequence.

Perhaps we can leave this endeavour for a moment. I want to show you a simple notation that is often used in connection with these ladder numbers, as you termed them. The notation makes it is less cumbersome to deal with the tiers that arise with continued fractions. It tames those unwieldy monsters.

Let's see it.

The two-tier fraction

$$1 + \frac{1}{2}$$

is written as [1;2] while the three-tier fraction

$$1 + \cfrac{1}{2 + \cfrac{1}{2}}$$

is written as [1;2,2]. This notation is a compact way of representing those awkward-to-write ladder numbers.

Let me see if I have grasped the notation. So [1;2,2,2] is shorthand for

$$1+\cfrac{1}{2+\cfrac{1}{2+\cfrac{1}{2}}}$$

—the four-tier ladder with its three 2s?

Exactly!

Why the semicolon after the first number?

The number 1 in front of the semicolon is the 1 standing out on its own before the first plus sign. It is the integer part of the number being represented. The numbers that come after the semicolon represent the tiered fraction part that comes after the first plus sign.

All those 2s down the side. How is the number 1 on its own written?

As [1], without a semicolon, it being understood that it is an integer.

Simplicity itself.

With this notation we can write the previous array of results as

$$\frac{1}{1}=[1]$$

$$\frac{3}{2}=[1;2]$$

$$\frac{7}{15}=[1;2,2]$$

$$\frac{17}{15}=[1;2,2,2]$$

—which I think you'll agree has its own appeal.

Yes.

And, were we to compute them, we'd find that

$$[1;2,2,2,2]=\frac{41}{29} \quad \text{and} \quad [1;2,2,2,2,2]=\frac{99}{70}$$

So in terms of the new notation, we want to show that the sequence

$$\frac{1}{1}, \frac{3}{2}, \frac{7}{5}, \frac{17}{12}, \frac{41}{29}, \frac{99}{70}, \ldots$$

may be written as

[1],   [1;2],   [1;2,2],   [1;2,2,2],   [1;2,2,2,2],   [1;2,2,2,2,2],...

which I bet is unlike any other sequence you have ever seen.

Have no doubts. I'd better set about proving that the terms of this unusual form of the sequence actually equal the corresponding terms of its more familiar form.

I don't expect you to have any trouble whatsoever.

With this new notation, the observation that

$$1+\cfrac{1}{2+\cfrac{1}{2+\cfrac{1}{2}}}=1+\cfrac{1}{1+\left(1+\cfrac{1}{2+\cfrac{1}{2}}\right)}$$

can be expressed as

$$[1;2,2,2]=1+\cfrac{1}{1+[1;2,2]}$$

which looks simple enough.

It does, but this new expression hides its sophistication. We are now at the stage where we must make that leap from the particular to the general, where arithmetic must give way to algebra. Just a little should do.

Let's hope I can find the right line of reasoning.

Don't worry about it; you already have.

I'm going to suppose $\frac{r}{s}$ is the fraction that results from computing one of these particular ladder numbers.

I see you have chosen two new letters to denote this fraction.

To avoid making any suggestions whatsoever.

New symbols for a new argument.

We are now expressing these ladder numbers in the form [1;2,2,2,...,2,2]. I put the ellipsis in the middle just to leave the number of 2s open.

Very sensible. You are getting the hang of this notation.

I'm going to assume that

$$[1;2,2,2,...,2,2]=\frac{r}{s}$$

— that the typical ladder number works out as $\frac{r}{s}$.

I'm with you.

Now I hope to show that the fraction resulting from the finite continued fraction consisting of just one more tier, or, in the new notation, the number $[1;2,2,2, \ldots , 2,2,2]$ with exactly one more 2 in it, is the fraction

$$\frac{r+2s}{r+s}$$

This then would be the one-step rule that generates the main sequence, expressed in terms of r and s rather than the customary m and n.

Yes.

And if you could do this, would you have your proof?

I believe so, because I know the first few ladder numbers reduce to the starting fractions of the sequence.

Point taken. So a ladder number having one more layer than one that reduces to a term in the sequence must reduce to the next term in the sequence.

That's the way I see it.

I'll let you get to the core of your proof.

Which is nothing more than that

$$[1;2,2,2,\ldots,2,2,2]=1+\cfrac{1}{1+[1;2,2,2,\ldots,2,2]}$$

based on how we know two successive ladder numbers are connected.

Magnificent!

The $[1;2,2,2, \ldots , 2,2]$ on the right-hand side is $\frac{r}{s}$, so the fraction representing the next ladder number after the one represented by $\frac{r}{s}$ is given by

$$1+\cfrac{1}{1+\cfrac{r}{s}}$$

Would you not agree?

Without hesitation.

Great. We have seen this expression before, with p where there is now an r, and q where there is now an s. Using the result of how this fraction simplified, we get

$$\frac{r+2s}{r+s}$$

in this instance.

Another great achievement done from start to finish. So now
we can say with certainty that the sequence of ladder numbers

$$1, \quad 1+\frac{1}{2}, \quad 1+\frac{1}{1+\frac{1}{2}}, \quad 1+\frac{1}{1+\frac{1}{1+\frac{1}{2}}}, \quad \cdots$$

is just the sequence of fractions

$$\frac{1}{1}, \quad \frac{3}{2}, \quad \frac{7}{5}, \quad \frac{17}{12} \cdots$$

in a different guise. These ladder numbers are called *finite continued fractions* for obvious reasons.

They are a rather peculiar way to represent "normal" fractions.

True. But when expressed in the more compact notation as

$$[1], \quad [1;2], \quad [1;2,2], \quad [1;2,2,2], \quad [1;2,2,2,2], \quad [1;2,2,2,2,2,2] \ldots$$

they have an intriguingly simple pattern to them.

I assume it is safe to say that, in this notation,

$$\sqrt{2} = [1; 2, 2, 2, 2, 2, \ldots]$$

with 2s all the way along.

Ad infinitum. The infinite continued fraction is often abbrevi-
ated to $[1; \bar{2}]$ with $\bar{2}$ indicating that the 2s continue forever and
ever.

Something like what is done for periodic decimal expansions.

Yes. We can write that

$$\sqrt{2} = [1; \bar{2}]$$

The almost alarmingly brief right-hand side encapsulating the
fact that every rung of the infinite ladder is the same.

Apart from the first.

As you say. This extreme simplicity makes $\sqrt{2}$ one of the most
stately within the realm of infinite continued fractions and is
the reason it possesses a certain property, which will be the
basis of our final excursion.

Much later, I hope.

Well, we still have a few things to discuss.

All Fractions

Each fraction in the main sequence has a finite continued fraction expansion.

That's right.

> Does every single fraction have a finite continued fraction expansion and, if so, how is it found?

Every fraction has a finite continued fraction expansion, but some expansions are easier to obtain than others.

> For example?

The finite continued fraction expansion of $\frac{20}{13}$ is obtained as follows:

$$\frac{20}{13} = 1 + \frac{7}{13}$$

$$= 1 + \cfrac{1}{\cfrac{13}{7}}$$

$$= 1 + \cfrac{1}{1 + \cfrac{6}{7}}$$

$$= 1 + \cfrac{1}{1 + \cfrac{1}{\cfrac{7}{6}}}$$

$$\Rightarrow \frac{20}{13} = 1 + \cfrac{1}{1 + \cfrac{1}{1 + \cfrac{1}{6}}}$$

This is a tedious enough calculation, but not overly long.

> You can say that again, but I think I see how it works.

Which is?

> You write the fractional portion that is less than 1 as 1 over its reciprocal, as you do when you write $\frac{6}{7}$ as 1 over $\frac{7}{6}$.

Yes.

> Next the fraction in the reciprocal that is bigger than 1 is written as an integer plus a new fraction less than 1. Then you do the same again until you can go no further.

That's it. Using the compact notation, we write that

$$\frac{20}{13} = [1;1,1,6]$$

Witness what we need to do when we simply change the numerator here from 20 to 21.

> I'm betting it's torture.

You might like to do the calculation and count the number of steps it takes.

I'll give it a go. I get

$$\frac{21}{13} = 1 + \frac{8}{13} = 1 + \cfrac{1}{\cfrac{13}{8}}$$

$$= 1 + \cfrac{1}{1 + \cfrac{5}{8}} = 1 + \cfrac{1}{1 + \cfrac{1}{\cfrac{8}{5}}}$$

$$= 1 + \cfrac{1}{1 + \cfrac{1}{1 + \cfrac{3}{5}}} = 1 + \cfrac{1}{1 + \cfrac{1}{1 + \cfrac{1}{\cfrac{5}{3}}}}$$

$$= 1 + \cfrac{1}{1 + \cfrac{1}{1 + \cfrac{1}{1 + \cfrac{2}{3}}}} = 1 + \cfrac{1}{1 + \cfrac{1}{1 + \cfrac{1}{1 + \cfrac{1}{\cfrac{3}{2}}}}}$$

$$\Rightarrow \frac{21}{13} = 1 + \cfrac{1}{1 + \cfrac{1}{1 + \cfrac{1}{1 + \cfrac{1}{1 + \cfrac{1}{2}}}}}$$

Whew! That took forever.

How many steps?

I count nine equal signs, so I'll say nine steps.

Even though it was hard going, it is impressive. In compact notation

$$\frac{21}{13} = [1;1,1,1,1,2]$$

it is all 1s except for the 2 at the end. We could have all 1s if we are happy to write

$$\frac{21}{13} = 1 + \cfrac{1}{1 + \cfrac{1}{1 + \cfrac{1}{1 + \cfrac{1}{1 + \cfrac{1}{1+1}}}}}$$

but it calls for a modification to the compact notation to record it.

I see. I presume the presence of all 1s is no accident.

They arise because the fraction I chose hails from a rather special family.

A story for another day, no doubt.

Yes. Here is a problem for you: How can we be sure that the continued fraction expansion of a normal fraction is finite?

Thinking about this should keep me quiet for a while.

Hero's Way

It must seem to you at this stage that we have been talking about nothing other than the sequence

$$\frac{1}{1}, \quad \frac{3}{2}, \quad \frac{7}{5}, \quad \frac{17}{12}, \quad \frac{41}{29}, \quad \frac{99}{70}, \quad \frac{239}{169}, \quad \frac{577}{408}, \dots$$

and matters directly related to it.

I've been surprised by the number of things there are to say about this sequence. I'm not sure I could recall them all. But I suppose one of the most important is that successive terms of this sequence provide better and better approximations of $\sqrt{2}$.

Without doubt, in the context of finding more and more of the digits in the decimal expansion of $\sqrt{2}$.

Something we haven't done for quite some time now.

I know. I have been conscious of this, wondering when you would take me to task about it.

It had slipped my mind until you mentioned decimal digits just now.

All of which I'll interpret as a sign that you have been enjoying our excursions, even if they haven't been exclusively focused on extending our knowledge of the decimal expansion of $\sqrt{2}$.

Besides enjoying myself, I have been thriving mentally.

Questions relating to this sequence made us consider two of its subsequences, as well as other related sequences. Furthermore, the identity

$$\sqrt{2} = 1 + \frac{1}{1 + \sqrt{2}}$$

allowed us see the sequence in a different light and introduced us to continued fractions, or ladder numbers, as you dubbed them. At every turn a new question or exploration always seemed to suggest itself.

Didn't they just?

It's the nature of any type of inquiry. If you are attracted to investigations of this kind, it is quite easy to understand how you could spend a good portion of your time answering and posing mathematical questions.

Which is what I'm told many mathematicians do.

Almost certainly.

Are we about to increase our knowledge of the decimal expansion of $\sqrt{2}$?

Well, we can always calculate as many terms of the sequence as we please, as we could have done at any time up to now, and just convert the last fraction to its decimal equivalent to obtain an approximation to $\sqrt{2}$. We don't have to do that much to obtain an approximation that is far in excess of the accuracy ever likely to be needed for any practical purpose.

But if we did that, what more would we have to talk about?

Some might say that we should have done this long ago and be done with it.

Never! I'm not one of them. Besides, I'm learning skills that I'm sure to be able to use elsewhere; and even if I don't, they're no burden.

I'm glad to hear it. With regard to your question about finding more digits of the decimal expansion of $\sqrt{2}$, I'd like to hold off on doing so for a while longer, and I hope you can bear with me on this point.

Of course. Are you building up to some sort of grand finale?

In a sense, yes. I want us to develop more powerful methods than that offered simply by using our main sequence. To begin, let me show you one way of finding another approximation method that might strike you as spectacular in comparison to what we already know.

Spectacular? This has to be interesting.

The method has been known since at least the first century, and many suspect that it was used well before that by the ancient Babylonians, whom we mentioned earlier in connection with their sexagesimal approximations of $\sqrt{2}$.

> Fascinating! Does this method use a completely different idea?

The derivation I'm about to show you is certainly different from anything we have done up to now. The method may have a surprise in store for you.

> A pleasant one, I hope.

We'll need to use a little algebra once again. I think it fair to say that the pieces of algebra used up to now, while being clever and often far from obvious, aren't overwhelmingly difficult to follow, particularly when you see them unfold in context.

> Will the algebra you are about to use be any more difficult than before?

I don't believe it will be.

> Let's get going then. I'm ready for action!

We begin by supposing that we have an approximation a to $\sqrt{2}$ and that we want to find a better one.

> You are using the letter a because it is the initial of the word "approximation"?

Yes; and it has no connection with any a we might have used before. As you know, it is better to use a letter, such as a, rather than an actual number when we want to discover a method that can be used in general.

> And could you not make such discoveries working with concrete numbers rather than symbols?

Maybe, but it is less likely. As I may have said before, sometimes you cannot see the forest for the trees when working with specific numbers.

> I have a better appreciation of this point now than I used to have. I'll let you get on with what you want to say.

Well, since a is only an approximation of $\sqrt{2}$, it is in error by some amount epsilon, ε, say. Mathematicians often use this character from the Greek alphabet to suggest a "small" quantity.

> It is also similar to e, the initial of error.

An added bonus. So let us write $a + \varepsilon = \sqrt{2}$.

> Are you assuming that a is an underestimate of $\sqrt{2}$, which would make ε positive?

A good question, to which the answer is no, even though the expression $a + \varepsilon = \sqrt{2}$ might seem to suggest this. If a is an overestimate, which it may be, then the quantity *epsilon* will actually be negative.

So the algebra you are about to do can handle both possibilities? Yes. If

$$a + \varepsilon = \sqrt{2}$$

then squaring both sides of this equation gives

$$(a + \varepsilon) \times (a + \varepsilon) = \sqrt{2} \times \sqrt{2}$$

or

$$a^2 + 2a\varepsilon + \varepsilon^2 = 2$$

$$
\begin{array}{l}
a + \quad \varepsilon \\
a + \quad \varepsilon \\
\hline
a^2 + \quad a\varepsilon \\
\quad + \quad a\varepsilon + \varepsilon^2 \\
\hline
a^2 + 2a\varepsilon + \varepsilon^2
\end{array}
$$

because of course $\sqrt{2} \times \sqrt{2} = 2$.

The fundamental relationship defining $\sqrt{2}$ used once again.

This equation can be viewed as a "quadratic" equation in ε, but we make no use of this perspective.

I remember quadratic equations from school, and that there is a formula that solves them all. You are not going to assume I know this formula and can use it, are you?

No, I am not. Besides it wouldn't lead us anywhere as it contains square roots.

That's good to hear. Are we trying to avoid square roots?

In a sense, yes, particularly ones we can't work out in terms of fractions. We are looking for approximations to one particular square root, namely $\sqrt{2}$, while not talking that much about the actual number itself. I'm going to explain the clever idea that is brought to bear on the above equation.

Which is?

To drop the ε^2 term. If ε is small, as we hope it is, then ε^2 is considerably smaller; so much so that we intend ignoring it. Dropping this "higher-order term" from the above equation gives

For example,
$(0.01)^2 = 0.0001$

Reminder: $\approx$ means
"Is approximately."

$$a^2 + 2a\varepsilon \approx 2$$

which is a much simpler expression than the previous equation and spares us the use of the quadratic formula with its undesirable square root.

But can you do this, simply drop a term?

Yes, we discard it without a thought! Very cheeky I know, and not what one expects of mathematicians who are supposed to be very precise and take everything into consideration.

My thoughts approximately!

Hmm. Sometimes bending the rules is just like taking a detour around an obstacle too hard to remove. This new expression is very convenient because it can be solved easily to give

$$\varepsilon \approx \frac{2-a^2}{2a}$$

Here ε is expressed in terms of the known approximation a.

What now?

Well, presumably—and we may not say more at this stage—adding this estimate of ε to a will provide us with an improved approximation of $\sqrt{2}$.

> So if a is an underestimate to begin with, then ε will turn out to be positive and improve this underestimate; whereas if a is an overestimate, ε will turn out to be negative, and adding it to a will bring the overestimate down, improving it also. Is that it?

Yes. Now we can say that

$$a+\varepsilon \approx a + \frac{2-a^2}{2a}$$
$$= \frac{2a^2+2-a^2}{2a}$$
$$= \frac{a^2+2}{2a}$$

Bringing the 2 in the denominator out in front as $\frac{1}{2}$, and dividing each term in the numerator by the a remaining in the denominator, we get

$$\frac{1}{2}\left[a+\frac{2}{a}\right]$$

as the concise expression for our new approximation of $\sqrt{2}$.

> With all trace of ε removed.

It was mere scaffolding. So

$$a \to \frac{1}{2}\left[a+\frac{2}{a}\right]$$

Read
$\to$
as "becomes".

is our new approximation rule.

> Looks simple enough.

Expressed in terms of one letter, which is the absolute minimum any formula can have.

> Will you elaborate a little on how this rule is used?

Of course. It says that if a is an approximation to $\sqrt{2}$, then another approximation is obtained by calculating the expression to the right of the $\rightarrow$.

I understand.

Why don't we become familiar with the rule by trying it on some numbers.

Okay. I'm going to start with $a = 1$.

Why $a = 1$?

Because it's the same as the seed $\frac{1}{1}$ in our main sequence.

And you are curious to compare the new recipe with the old one?

More than curious.

An excellent idea; full speed ahead!

Putting $a = 1$ into the right-hand side gives

$$\frac{1}{2}\left[1+\frac{2}{1}\right]=\frac{1}{2}[3]=\frac{3}{2}$$

This is the second fraction in our original sequence.

Are you disappointed?

Not really. At least it is an improvement on the initial approximation of 1. I'm now going to update a to have the value $\frac{3}{2}$ to see what pops out next time round.

A good idea.

With $a = \frac{3}{2}$, we get

$$\frac{1}{2}\left[\frac{3}{2}+\frac{2}{\dfrac{3}{2}}\right]=\frac{1}{2}\left[\frac{3}{2}+\frac{4}{3}\right]=\frac{17}{12}$$

which is a most welcome new wrinkle.

A welcome new wrinkle—what do you mean?

New, because the method produces $\frac{17}{12}$ rather than the $\frac{7}{5}$, which is the next term after $\frac{3}{2}$ in the sequence.

$$\frac{1}{1},\ \frac{3}{2},\ \frac{7}{5},\ \frac{17}{12},\ \frac{41}{29},\ \frac{99}{70},\ \frac{239}{169},\ \frac{577}{408},\ \dots$$

And welcome because $\frac{17}{12}$ is a better approximation of $\sqrt{2}$ than $\frac{3}{2}$ or $\frac{7}{5}$; and a wrinkle because $\frac{17}{12}$ is itself another fraction from the sequence, which is a nice surprise.

Yes, this was the surprise I promised earlier.

> You mean that the new method appears to be generating fractions that are in the sequence.

Yes.

> I'm not saying that I understand why, but the fact that I chose $a = 1$ as the starting approximation must have something to do with it.

Certainly. Had you not chosen it, we may not have got terms that are part of the sequence.

> I'm glad I obliged. I suppose I could show pretty quickly that a starting value of a different from any fraction appearing in this sequence generates an entirely different sequence of fractions.

Later, when you get a moment to yourself, try the seed $\frac{2}{1}$, which is not in the fundamental sequence, to see what you get.

> Will do. This new method just skips over $\frac{7}{5}$ to land on $\frac{17}{12}$.

It does.

> I'm going to see if it does more skipping by updating a to $\frac{17}{12}$.

Exactly what I would do myself.

> All right then. With $a = \frac{17}{12}$, we get

$$\frac{1}{2}\left[\frac{17}{12} + \frac{2}{\frac{17}{12}}\right] = \frac{1}{2}\left[\frac{17}{12} + \frac{24}{17}\right] = \frac{577}{408}$$

> —the last of the eight terms that we worked out in our sequence!

This time the method skipped three fractions before landing on the eighth term.

> This new method is spectacular, as you said. It's more powerful.

More powerful in what sense?

> In producing successive approximations that get closer and closer to $\sqrt{2}$ much more quickly.

It certainly seems so.

> Starting with $a = 1 = \frac{1}{1}$, it generates a subsequence whose leading terms are

$$\frac{1}{1}, \quad \frac{3}{2}, \quad \frac{17}{12}, \quad \frac{577}{408}, \ldots$$

Since these fractions occupy positions 1, 2, 4, and 8 in the fundamental sequence, it seems to be generating a very special subsequence of the main sequence.

A Little History

Before we do anything else I must tell you that these fractional approximations to $\sqrt{2}$ are believed to be the precise ones the Babylonians used over three thousand years ago.

> You did mention before that some of these fractions were known to the Babylonians. So did they know this method then?

Well, a more general version of the rule we have just developed, which finds approximations to any square root, is credited to Hero of Alexandria in the first century A.D.

> The first century A.D. is an awfully long time after 1600 B.C. which is roughly the time you said that the Babylonians first knew of these approximations.

Yes, it is intriguing. Did they know Hero's method all those years before him? If so, as some believe, then it must be in the running for the title of, as someone once put it, "The world's oldest algorithm."

> How did they, or Hero, come up with the method—the same way we did?

One theory is that they knew that if a is a positive approximation on one side of $\sqrt{2}$ on the number line, then $\frac{2}{a}$ is on the opposite side of $\sqrt{2}$. Since the average of two numbers lies between the two numbers, this average must be closer to $\sqrt{2}$ than either a or $\frac{2}{a}$, and so is an improvement on either of them as an approximation of $\sqrt{2}$.

> Which settles the improvement question straight off. And this average is none other than

$$\frac{1}{2}\left[a+\frac{2}{a}\right]$$

> Brilliant, and so simple.

A gem!

> Is it difficult to prove that if a is a positive approximation on one side of $\sqrt{2}$, then $\frac{2}{a}$ is on the opposite side of $\sqrt{2}$?

Simple; as I'll show you. Afterward you might like to work up a numerical example.

> It's a bargain.

If $a < \sqrt{2}$, then multiplying both sides by $\sqrt{2}$ gives $\sqrt{2}a < 2$. Dividing this inequality through by a preserves the inequality sign since a is positive. We get

$$a<\sqrt{2}\Rightarrow\sqrt{2}<\frac{2}{a}$$

<
means
"is less than".

Exactly the same argument works with the inequality signs reversed to give

$$a > \sqrt{2} \Rightarrow \frac{2}{a} < \sqrt{2}$$

So whichever side of $\sqrt{2}$ the approximation a is on, the quantity $\frac{2}{a}$ is on the other side.

Easy when you know how! For my numerical example I've taken a to be the fraction $\frac{7}{5}$ from the main sequence. As we know, it is less than $\sqrt{2}$. Then

$$\frac{7}{5} < \sqrt{2} \Rightarrow \frac{2}{\frac{7}{5}} \Rightarrow \frac{10}{7} > \sqrt{2}$$

Let me check this.

$$\frac{100}{49} = \frac{98+2}{49} = 2 + \frac{2}{49} \Rightarrow \frac{100}{49} > 2 \Rightarrow \frac{10}{7} > \sqrt{2}$$

As asserted.

In this case

$$\frac{1}{2}\left(a + \frac{2}{a}\right) = \frac{1}{2}\left(\frac{7}{5} + \frac{10}{7}\right) = \frac{99}{70}$$

Another fraction from the main sequence. Do you want me to do a numerical example with $a > \sqrt{2}$?

Not at all. Let us return to where we were before taking this historical detour.

The Heron Sequence

I have a few questions about Hero's rule and the sequence

$$\frac{1}{1}, \frac{3}{2}, \frac{17}{12}, \frac{577}{408}, \ldots$$

generated by this rule.

Some historical sources refer to Hero as Heron, so we might refer to this sequence as the Heron sequence.

Saves having to write down the sequence every time we want to talk about it.

Or, if we want, we could say the Heron sequence with seed 1.

Because beginning Hero's method on a different seed will generate a different Heron sequence?

Different in its first term, at the very least. But let's just call this sequence the Heron sequence, and if we need to talk about another sequence generated by the same rule, we'll take care to mention its seed.

> So when we say the Heron sequence we mean this one with seed $\frac{1}{1}$?

I think we'll allow ourselves this little luxury for the sake of brevity. So, what questions do you have?

> Firstly, I want to know if the next term after $\frac{577}{408}$ in the Heron sequence is the sixteenth from the fundamental sequence with the next one after that being the thirty-second, and so on?

With the position of each successive term being double that of its predecessor's in the fundamental sequence. A very fine question. I notice that you are now referring to our original, or main, sequence as the fundamental sequence.

> I did, didn't I? You used this term earlier, if I'm not mistaken.

I did indeed; which is fine because it is fundamental in its connection with $\sqrt{2}$.

> Anyway, I'm also curious to know if all the fractions after the first in the Heron sequence are overestimates of $\sqrt{2}$.

What makes you think this?

> Because, except for the seed, which is an underestimate of $\sqrt{2}$, the other three fractions that we have worked out are all in even positions in the fundamental sequence.

And so belong to its over-subsequence, as we termed it. Nicely observed—the terms of the subsequence consisting of every second term from the fundamental sequence are precisely those that overestimate $\sqrt{2}$.

> If I could nail my first conjecture, I'd have the second one for free.

It would seem that way, since 2, 4, 8, 16, 32 . . . are all even numbers. But are you sure that the Heron sequence is a subsequence of the main sequence?

> This is a twist. I didn't think about it. Surely it is?

Based on what evidence—a few terms agreeing? What would a barrister say to you?

> Oh, what a pain! Why can't life be simple?

I'm sure you'll think of something.

> Not this time. To me, Hero's rule looks completely different from our original rule. I'm definitely going to need help.

What's one sure way of knowing whether a fraction is in the main sequence or not?

Hard-thinking time again. I remember now. We proved that if a fraction has the property that its numerator squared minus twice its denominator squared is either 1 or −1, then it is a term in the fundamental sequence.

Yes, we did, in what was one of our more strenuous excursions. So?

I'm glad we have this result under our belts because I can see now that it is vital.

Why?

If I can show that the fractions of the Heron sequence have this property, then I could convince that barrister fellow that they are from the fundamental sequence.

And you'd win your case.

This particular one at any rate. So how do I show that the Heron fractions are fundamental fractions?

Fundamental fraction—another nice term. By discussing the typical fraction in the Heron sequence.

Since this is another new investigation, I know I should use a new pair of letters for the numerator and denominator of this typical fraction.

It is not sacrosanct that you use fresh symbols in every new situation. You can use any letters you like, as long as you are clear about what they represent.

In that case, I'll go back to $\frac{m}{n}$.

Just keep in mind that it is Hero's rule that is being applied to it and that it is a typical member of the Heron sequence.

Will do. So to begin I must replace a by $\frac{m}{n}$ in

$$\frac{1}{2}\left[a+\frac{2}{a}\right]$$

Is that right?

A good place to start.

Now let me see if my algebraic manipulation can manage this and tidy everything up afterward. I get

$$\frac{1}{2}\left[\frac{m}{n}+\frac{2}{\frac{m}{n}}\right]=\frac{1}{2}\left[\frac{m}{n}+\frac{2n}{m}\right]$$

$$=\frac{1}{2}\left[\frac{m^2+2n^2}{nm}\right]$$

$$=\frac{m^2+2n^2}{2mn}$$

Very ably handled.

This is very different in appearance from our previous rule.

Noticeably. A different rule for a different type of sequence.

So the rule for generating successive terms of the Heron sequence is

$$\frac{m}{n} \to \frac{m^2 + 2n^2}{2mn}$$

It isn't as simple-looking as the rule

$$\frac{m}{n} \to \frac{m+2n}{m+n}$$

which generates the fundamental sequence from the seed $\frac{1}{1}$.

Agreed. It says that the numerator of the fraction after $\frac{m}{n}$ is $m^2 + 2n^2$, while its denominator is $2mn$. It involves the squares of m and n, and even a multiplication of m by n, whereas the previous rule involved only additions and one simple multiplication by 2.

When you put it like this, it sounds as if it's a complicated rule.

Anyway, your hope is to show that it generates a subsequence of the original sequence when applied to the seed $\frac{1}{1}$.

Yes, that's my task at the moment.

This subsequence provides successive approximations of $\sqrt{2}$ that approach it more rapidly than do the terms of the full sequence generated by the simpler rule. We could say that the price for extra power is a more involved rule, although this Heron rule is not difficult to apply. Why don't you run a quick check on the rule in the case $\frac{m}{n} = \frac{1}{1}$ to see if it gives the fraction $\frac{3}{2}$?

Right. Setting $m = 1$ and $n = 1$ tells me that $m^2 = 1$ with $2n^2 = 2(1) = 2$. Thus $m^2 + 2n^2 = 1 + 2 = 3$.

As it should.

The new denominator is $2mn = 2(1)(1) = 2$, which is correct also. So

$$\frac{1}{1} \to \frac{3}{2}$$

The rule generates $\frac{3}{2}$.

Now convince me that the Heron sequence is a subsequence of the fundamental sequence.

If this new sequence is a subsequence of the original sequence, then its $(\text{top})^2 - 2(\text{bottom})^2$ must be either -1 or 1 also.

This is the key. Show this and you are done, because in this case, and in this case only, is the fraction a member of the fundamental sequence.

So, fingers crossed. I'll work out the square of the numerator first. Now

$$(m^2 + 2n^2)^2 = m^4 + 4m^2n^2 + 4n^4$$

if I've done this correctly.

Faultlessly.

I must now work out twice the square of the denominator. It is given by

$$2(2mn)^2 = 2(4m^2n^2) = 8m^2n^2$$

Now subtract this quantity from the previous quantity to obtain the difference between the square of the numerator and twice the square of its denominator.

Of course—this is the next step.

We get

$$(m^4 + 4m^2n^2 + 4n^4) - 8m^2n^2 = m^4 - 4m^2n^2 + 4n^4$$

This result is by no means as simple as I was hoping it would be.

It is a little intimidating, and not very revealing as it stands.

I'm glad you said that.

But can you spot another way to write $m^4 - 4m^2n^2 + 4n^4$?

Do you mean can I factorize it? It took all my limited skill to multiply out all the expressions with m's and n's everywhere. I wouldn't trust myself to factorize.

Well,

$$m^4 - 4m^2n^2 + 4n^4 = (m^2 - 2n^2)^2$$

which you should find very helpful.

What's so helpful, I wonder, about this $(m^2 - 2n^2)^2$, which is the square of the quantity $m^2 - 2n^2$?

Doesn't this say that the new difference, $m^4 - 4m^2n^2 + 4n^4$, if I may call it that, is just the old difference, $m^2 - 2n^2$, squared? Now finish off the argument.

But the old one is either -1 or 1 all the time. This means that the new difference must be 1. I have it!

You have. In fact, there can only ever be at most one -1, which there is in this case because of the seed $\frac{1}{1}$.

Because $(-1)^2 = 1$, all the others must be 1.

(side calculation, right margin)

$$\begin{array}{r} m + \ 2n \\ m + \ 2n \\ \hline m^2 + 2mn \\ + \ 2mn + 4n^2 \\ \hline m^2 + 4mn + 4n^2 \end{array}$$

Correct.

So we have actually shown two things, although we began by trying to show only one.

Care to elaborate?

Well, besides showing that the Heron sequence is a subsequence of the main sequence, we now understand why all its terms beyond the first are overestimates.

Why?

Because they all have a signature of 1, which we said earlier is the fast way of knowing that a fraction is an overestimate.

Well done again.

It is really satisfying to prove something in a simple way once you see how to go about it.

A real thrill indeed. The reward of understanding something that previously appeared mysterious is wonderful.

Speed and Acceleration

I still have to tackle the question about the rate at which the Heron rule "travels" along the fundamental sequence starting from $\frac{1}{1}$.

$$\frac{m}{n} \rightarrow \frac{m^2 + 2n^2}{2mm}$$

Hopping, as if it were, from the first term to the second, then onto the fourth, then to the eighth and so on, doubling its previous hop with each successive stride.

Exactly.

In marked contrast to the one-step rule

$$\frac{m}{n} \rightarrow \frac{m+2n}{m+n}$$

which strolls along from term to term, never altering its pace. The Heron rule accelerates while the one-step rule ambles long at a constant speed.

That's a nice way of looking at it.

Imagery that we will use from time to time. In which category would you place the rule

$$\frac{m}{n} \rightarrow \frac{3m+4n}{2m+3n}$$

—proceeding at an accelerated pace or traveling along at a fixed speed?

> It has been a while since we talked about this rule. If it starts at the seed $\frac{1}{1}$, it takes us through the fundamental sequence two steps at a time.

Yes, picking out all the odd-numbered underestimating fractions and skipping over all the even-numbered overestimating fractions.

> Start it at $\frac{3}{2}$ and it does the exact opposite.

True. But another plodder, it has to be said. Still, I suppose it has twice the speed of the one-step rule.

> As you say.

And as we said before, it is very handy if for some reason we want to generate the full under- or over-subsequences.

> I suppose there are rules that allow us to pick out every third term, or every fourth term and so on, of the fundamental sequence?

There are. We'll talk a little about them at a later stage. Of course, they must all be constant-speed rules.

> You'll have to forgive me, but I just have to use the Heron rule to get the next fraction in the Heron sequence after $\frac{577}{408}$.

If you must, you must.

> With $m = 577$ and $n = 408$, the approximation formula gives

$$\frac{m^2 + 2n^2}{2mn} = \frac{577^2 + 2(408^2)}{2(577)(408)} = \frac{665857}{470832}$$

and

$$(665857)^2 - 2(470832)^2 = 443365544449 - 443365544448 = 1$$

> Which I knew had to be the case.

But you just had to check, didn't you?

> Yes, I admit, but how did you know?

It's a fairly common behavior: no matter how much people believe the theory, they still like to see it in action on concrete numbers.

> By the way, is this latest arrival the sixteenth member of the fundamental sequence?

It is. We really must get round to displaying more than just the first eight terms of the fundamental sequence. Speaking of which, let me add this most recent acquisition to our existing list of known Heron fractions to get

$$\frac{1}{1}, \quad \frac{3}{2}, \quad \frac{17}{12}, \quad \frac{577}{408}, \quad \frac{665857}{470832}, \dots$$

as the latest update on this wonderful sequence.

The numerator and denominator of the last fraction have six digits each.

A solid citizen.

It's intriguing that these fractions would appear to have nothing in common because they look so different from each other, yet we know that they are actually almost right beside each other on the number line.

That's true, particularly of the later ones.

Time Out for a Sneak Preview

I know you intend to punch out a lot of digits of the decimal expansion of $\sqrt{2}$ sometime in the future, but could we take time out to estimate how many digits of the decimal expansion of the fraction

$$\frac{665857}{470832} = 1.4142135623746899106\dots$$

coincide with the leading digits of the decimal expansion of $\sqrt{2}$?

By all means. How do you want to go about it?

I thought that I'd get the fraction just after this one in the fundamental sequence with the one-step rule and then get the decimal expansions of both fractions. Whichever leading digits are common to both will be the leading digits in the expansion of $\sqrt{2}$.

Simple and effective. Though we'll get these decimal expansions using a computer; it is not cheating because, if we had to, we could work them out by hand.

So we won't be breaking our desert-island must-be-able-to-do-it-yourself code of conduct?

We won't. What is the fraction after $\frac{665857}{470832}$?

It is given by

$$\frac{665857 + 2(470832)}{665857 + 470832} = \frac{1607521}{1136689}$$

Very well. What now?

Well, since the Heron fraction is an overestimate, this one is an underestimate, and so

$$\frac{1607521}{1136689} < \sqrt{2} < \frac{665857}{470832}$$

Now I'll just get the decimal expansions of the fractions to a goodly number of decimal places, twenty say.

That should be plenty.

We get

$$1.41421356237282141377\ldots < \sqrt{2} < 1.41421356237468991063\ldots$$

which shows that

$$\sqrt{2} = 1.41421356237\ldots$$

to eleven decimal places.

Yes, and all this with just the fifth term in the Heron sequence and its successor in the fundamental sequence.

Always Over

Is it true that if we choose any fraction as an initial approximation to $\sqrt{2}$ then the second fraction produced by Hero's rule is always an overestimate?

Why do you ask this question?

Well, I was doing a bit of experimenting of my own on the "hopping conjecture"—you know my 1, 2, 4, 8, 16, . . . idea about the Heron sequence.

What exactly did you do?

I tried the Heron rule on the seed $\frac{3}{2}$, which I thought afterward was a little stupid because it's bound to generate just the Heron sequence without its seed of $\frac{1}{1}$. So then I tried the seed $\frac{7}{5}$ and got the sequence

$$\frac{7}{5}, \quad \frac{99}{70}, \quad \frac{19601}{13860}, \quad \frac{768398401}{543339720}, \quad \ldots$$

which I generated using a calculator.

Good for you.

These fractions are in positions 3, 6, 12, and 24 in the fundamental sequence. As you can see, they fit the doubling pattern. It was then I thought that if this doubling pattern is a feature of the Heron rule, it doesn't matter whether we start at an odd or even position, because once you double you're on even numbers from then on. All the fractions living in the even-numbered positions are overestimates.

Agreed, and very nice when you take a seed anywhere from the fundamental sequence. But what if you use a different seed?

I'm afraid I overlooked that detail. I'll have to think about the matter some more.

What you believe to be true is in fact the case.

Is it easy to show?

It is easy to follow the steps once they are shown to you, but the argument is very "slick," a word of praise when used about a proof.

I must see it, then.

A proof such as this one is found in practice by assuming that what you want to show is true, examining the implications of this assumption until you reach—if you are lucky—something that is self-evident. Then you try to work back from this obvious fact, and if your luck still holds, you succeed in reversing all the implications and arrive at the result you suspected to be true in the first place.

I'm not sure I follow all of this, but I get the gist. Do you have time to elaborate?

Always. Our suspicion is that if a is a positive approximation to $\sqrt{2}$, then

$$\frac{1}{2}\left(a+\frac{2}{a}\right)>\sqrt{2}$$

no matter what the value of a. Are you with me?

I think so.

Because a is positive, the terms on both sides of this inequality are positive. So when I square both sides to "get rid of $\sqrt{2}$," as is said, the inequality sign will still have the same direction.

A point you made to me earlier.

Doing this and tidying up a little gives

$$a^2+4+\frac{4}{a^2}>8$$

Check this at your leisure.

I think I can already see that it's true.

Good. When we subtract the 8 from the 4 and multiply the equation through by a^2, which is positive, the inequality sign is still preserved, so we get that $a^4-4a^2+4>0$. Now we write

$$a^4-4a^2+4>0 \quad \text{as} \quad (a^2-2)^2>0$$

Reminder:
> means
"is greater than."

This is the clever part.

I'll check all these technical steps later. Have we arrived at something that is obviously true?

Yes. Since a is only an approximation of $\sqrt{2}$, it is not equal to $\sqrt{2}$. Thus, $a^2 \neq 2$ and so $a^2 - 2 \neq 0$.

$\neq$ means "not equal to."

I think I follow this.

Now what happens when a nonzero quantity is squared?

It's always positive.

Exactly, whether it be positive or negative, its square is always positive. So $(a^2 - 2)^2 > 0$. It is with this fact that we start our proof.

You are going to reverse all the steps?

If they are reversible. I'll streamline the whole argument. Are you ready?

As I ever will be.

If $a > 0$ is an approximation of $\sqrt{2}$, then

$$\left(a^2 - 2\right)^2 > 0 \Rightarrow a^4 - 4a^2 + 4 > 0$$

$$\Rightarrow a^2 - 4 + \frac{4}{a^2} > 0 \quad \text{(because } a^2 > 0\text{)}$$

$$\Rightarrow a^2 + 4 + \frac{4}{a^2} > 8$$

$$\Rightarrow \left(a + \frac{2}{a}\right)^2 > 8$$

$$\Rightarrow a + \frac{2}{a} > 2\sqrt{2} \quad \text{(square roots of positives)}$$

$$\Rightarrow \frac{1}{2}\left(a + \frac{2}{a}\right) > \sqrt{2}$$

shows that the quantity

$$\frac{1}{2}\left(a + \frac{2}{a}\right)$$

is always greater than $\sqrt{2}$.

It must take years of practice to become this slick.

It does take time.

Isn't it rather amazing that the second approximation and all later ones are always greater than $\sqrt{2}$?

Rather like the second approximation obtained with our first rule always being between 1 and 2.

To Go Under

If the Heron method always provides overestimates of $\sqrt{2}$ after, at most, the first approximation, does it mean that the rule cannot be used to produce an accelerated sequence of underestimates? That would be disappointing.

It would.

It strikes me as a problem that we should be able to solve easily, but I can't quite see how to go about it.

Well, you used an idea a while back that holds the solution to this problem.

I did?

You did. Let us view this problem then as no more than a technical one, challenging us to put our theory into practice. Just think like an engineer seeking to chain different mechanisms.

Which ones? I don't quite follow.

Well, we know how to generate overestimates with the Heron rule, and we also know how to move back or forward one term to obtain an underestimate.

Of course we do. I used this idea when getting a decimal approximation to $\sqrt{2}$ using the fifth fraction in the Heron sequence. So if all else fails, we can get the computer to print the overestimates by Heron's method and we'll calculate the underestimates by hand. I'm joking!

A solution for those who are gluttons for punishment. We need to find a mechanism that generates an overestimating fraction and follows this by calculating its successor.

More algebraic manipulation.

Without question, but nothing to be frightened of. Think this idea through abstractly on the algebraic fraction $\frac{m}{n}$, and we'll solve the problem.

First, the Heron rule says that

$$\frac{m}{n} \to \frac{m^2 + 2n^2}{2mn}$$

Yes, and we know this latter fraction is an overestimate. So what do you need to do now?

Step it down to an underestimate.

How?

By either stepping backward or going forward.

Forward may be easier. Step forward to go under.

But how?

How do we get the next fraction in the fundamental sequence? Just think of the verbal rule you came up with and apply its instructions to this fraction.

In words, the new denominator is the old numerator added to the old denominator.

Translate this into symbols in the present context.

The new denominator is $(m^2 + 2n^2) + 2mn$.

Exactly, or $m^2 + 2mn + 2n^2$, as such expressions are often written to adhere to lexicographic ordering. New numerator?

The new numerator is the old numerator added to twice the old denominator. This translates to $(m^2 + 2n^2) + 2(2mn)$.

And simplifies to $m^2 + 4mn + 2n^2$.

So is the mechanism we want just the rule that turns the fraction $\frac{m}{n}$ into this new fraction?

None other. Write it down and then test it to see it work like a charm.

I can't wait. The "under-rule" is

$$\frac{m}{n} \rightarrow \frac{m^2 + 4mn + 2n^2}{m^2 + 2mn + 2n^2}$$

if we have gone about our business correctly.

Try it on the seed $\frac{1}{1}$ to get a succession of rapidly improving underestimates.

We get

$$1, \quad \frac{7}{5}, \quad \frac{239}{169}, \quad \frac{275807}{195025}, \quad \frac{367296043199}{259717522849}, \ldots$$

which are terms 1, 3, 7, 15, and I suppose term 31, in the fundamental sequence.

And are these term-numbers what we'd expect?

I think so. Without the step forward we'd get term 2, but stepped forward this goes to $2 + 1 = 3$. Then the doubling brings us to position 6, which when advanced by 1, gives the term from position 7 in the main sequence.

Then $2 \times 7 + 1 = 15$, with $2 \times 15 + 1 = 31$, and so on.

And since these are odd-numbered positions, they all represent underestimates.

Very pleasing. In fact, magnificently successful, I would say.

It's great how knowing the theory helped us solve the problem in hand.

Understand the theory and you can drive the practice.

Different Seed, Same Breed

I forgot to tell you what happened when I tried

$$\frac{m}{n} \to \frac{m^2 + 2n^2}{2mn}$$

on the seed $\frac{2}{1}$.

You joined the Heron sequence at position 2, in the person of the right-honorable $\frac{3}{2}$.

I did, and of course you knew I would. But I got quite a surprise. I never expected a different seed to breed the same future generations. Does this happen in other ways?

You mean can you tap into the Heron sequence at different points starting with other seeds, say at the third position or fourth or any other position for that matter?

Something like that.

Yes. If you apply the Heron rule to the fraction $\frac{2n}{m}$, you also get . . . why don't you work it out to see, and then you'll have investigated everything thoroughly.

But I still haven't gotten anywhere with my investigation concerning the doubling phenomenon of the Heron rule.

Perhaps our final piece of witchcraft will shed some light on this burning question.

All in the Family

Are you about to develop rules that provide better approximations of $\sqrt{2}$ at an even faster rate than Hero's rule?

Yes. We're going to speed up the pace by steadily increasing the pressure on the accelerator, as it were.

Can you not put the pedal to the metal?

You'll be able to answer this question for yourself in a little while.

We're in for some fast action then? I'm expecting great things, considering how impressive the Heron sequence is.

Perhaps you shouldn't set your expectations too high. You might find the whole business a bit of a letdown.

I'm sure I won't.

If $\frac{m}{n}$ and $\frac{p}{q}$ are any two fractions in the fundamental sequence

$$\frac{1}{1}, \frac{3}{2}, \frac{7}{5}, \frac{17}{12}, \frac{41}{29}, \frac{99}{70}, \frac{239}{169}, \frac{577}{408}, \ldots$$

then we know that

$$m^2 - 2n^2 = \pm 1 \quad \text{and} \quad p^2 - 2q^2 = \pm 1$$

where, as usual, ± 1 means either 1 or -1.

And $\frac{m}{n}$ and $\frac{p}{q}$ each have stood for typical fractions in this sequence on a number of occasions.

Indeed they have. Now we are about to see them trip the light fantastic together.

Some more of the witchcraft you mentioned a while back.

I hope you'll think so. The first piece of conjuring is to multiply the expression $m^2 - 2n^2 = \pm 1$ by $p^2 - 2q^2 = \pm 1$ to get

$$(m^2 - 2n^2)(p^2 - 2q^2) = \pm 1$$

Can you explain why we get ± 1 on the right-hand side of this equation?

Let me see. Each ± 1 stands for either 1 or -1, so when multiplied by each other, the answer must come out to be either 1 or -1.

That's it exactly. We are going to mine this equation deeply by way of some pyrotechnics.

Entertainment!

We aim to please. Now what I'm about to do is change this product

$$(m^2 - 2n^2)(p^2 - 2q^2)$$

into a single term using the defining relation for $\sqrt{2}$, namely that $2 = \sqrt{2} \times \sqrt{2}$.

This is normally written the other way around, isn't it?

I write it in reverse for a reason. With $2 = \sqrt{2} \times \sqrt{2} = (\sqrt{2})^2$,

$$m^2 - 2n^2 = m^2 - (\sqrt{2}n)^2 = (m - \sqrt{2}n)(m + \sqrt{2}n)$$

Difference of two squares

The right-hand side is a factorization of the $m^2 - 2n^2$ on the left-hand side. This is the first time we have used this maneuver.

I was never that comfortable with factorization at school.

$$\begin{array}{r} m - \ \sqrt{2}n \\ m + \ \sqrt{2}n \\ \hline m^2 - \sqrt{2}mn \\ + \ \sqrt{2}mn - 4n^2 \\ \hline m^2 + \quad 0 - 4n^2 \end{array}$$

Don't worry, the job is done. You might, when you have time, multiply the two terms on the right-hand side to verify that the left-hand side is obtained.

Multiplying I think I can manage, it's undoing the multiplication I don't warm to.

Be sure to note where the relationship defining $\sqrt{2}$ is used.

Will do.

Similarly, $p^2 + 2q^2 = (p - \sqrt{2}q)(p + \sqrt{2}q)$. Now all is in readiness for more fireworks.

Are we in for some tough going?

Yes, but we'll take it slowly—and the result will be worth the effort. To begin

$$(m^2 - 2n^2)(p^2 - 2q^2) = [(m - \sqrt{2}n)(m + \sqrt{2}n)][(p - \sqrt{2}q)(p + \sqrt{2}q)]$$
$$= [(m - \sqrt{2}n)(p - \sqrt{2}q)][(m + \sqrt{2}n)(p + \sqrt{2}q)]$$

In the second line, the two expressions with the minus sign in their middles are paired between a set of square brackets, as are the two with the plus sign in their middles.

I assume you have your reasons for doing this.

Which will become clear soon, I hope. Now, a little multiplication shows that the first pair simplifies to $(mp + 2nq) - \sqrt{2}(mq + np)$, while the second pair simplifies to $(mp + 2nq) + \sqrt{2}(mq + np)$.

I'll take your word for it.

Yes, these multi-lettered products are the same except for the opposite signs in the middle. You can also take my word for it that

$$[(mp + 2nq) - \sqrt{2}(mq + mp)][(mp + 2nq) + \sqrt{2}(mq + mp)]$$
$$= (mp + 2nq)^2 - 2(mq + np)^2$$

This calculation is of the form $(a - \sqrt{2}b)(a + \sqrt{2}b) = a^2 - 2b^2$.

Where $a = mp + 2nq$ and $b = mq + np$?

Yes.

Whew! This is pretty hairy stuff from where I'm standing.

I know, because of those off-putting letters. But it's all just scaffolding, which we now discard to reveal the identity

$$(m^2 - 2n^2)(p^2 - 2q^2) = (mp + 2nq)^2 - 2(mq + np)^2$$

which is true for *all* values of *m*, *n*, *p* and *q*, no matter where they hail from.

I'm afraid at the moment I feel more overwhelmed than impressed.

I understand; you are probably reeling. Anyway, how can you be impressed about this relationship, which I have dignified by calling an identity, when you don't yet know the destination of this mystery tour. For now I'll be quite happy if you accept that it is true and watch how it is put to work.

> Well, if it's any consolation, I can see that the two terms multiplying each other on the left become a single term on the right, which was your objective.

Indeed. And can you see that the term on the right has exactly the same form as each of the terms on the left?

> In the sense that, like them, it is something squared minus twice something else squared?

Exactly. Now, let us return to our sequence. As we have already said, if $\frac{m}{n}$ and $\frac{p}{q}$ are any fractions from the sequence

$$\frac{1}{1}, \quad \frac{3}{2}, \quad \frac{7}{5}, \quad \frac{17}{12}, \quad \frac{41}{29}, \quad \frac{99}{70}, \quad \frac{239}{169}, \quad \frac{577}{408}, \ldots$$

then

$$(m^2 - 2n^2)(p^2 - 2q^2) = \pm 1$$

Because of our newly acquired identity, this means that

$$(mp + 2nq)^2 - 2(mq + np)^2 = \pm 1$$

What does this tell us?

> That the fraction

$$\frac{mp + 2nq}{mq + np}$$

> is also a member of the sequence.

It does indeed; well spotted. But explain why anyway.

> Because we proved that if a fraction has the property that its numerator squared minus twice its denominator squared is either 1 or −1, then it is a term in the fundamental sequence. We used this result at least once before.

True. It is very important. A fraction with this property is a fundamental fraction. Let's make clear what it is we have achieved up to this point in time.

> Does this mean that the merry dance between $\frac{m}{n}$ and $\frac{p}{q}$ has ended?

Nearly. When you think about what we have just done, you see that the two fundamental fractions $\frac{m}{n}$ and $\frac{p}{q}$ can be combined to produce another fraction, namely

$$\frac{mp+2nq}{mq+np}$$

which is also in the fundamental sequence.

This is very interesting; another kind of rule.

Try some examples to get better acquainted with what's afoot.

I'll start out with the first and second terms

$$\frac{m}{n}=\frac{1}{1} \quad \text{and} \quad \frac{p}{q}=\frac{3}{2}$$

Taking $m = 1$, $n = 1$, $p = 3$ and $q = 2$ makes

$$\frac{mp+2nq}{mq+np}=\frac{(1\times3)+2(1\times2)}{(1\times2)+(1\times3)}=\frac{7}{5}$$

— the next and third term in the sequence. So the first and second combine to give the third. Just $1 + 2 = 3$.

Why don't you try the new combination rule again, this time with

$$\frac{m}{n}=\frac{3}{2} \quad \text{and} \quad \frac{p}{q}=\frac{7}{5}$$

so that you can take $m = 3$, $n = 2$, $p = 7$ and $q = 5$.

The second term combined with the third. We get

$$\frac{mp+2nq}{mq+np}=\frac{(3\times7)+2(2\times5)}{(3\times5)+(2\times7)}=\frac{41}{29}$$

We have skipped over $\frac{17}{12}$ and arrived at $\frac{41}{29}$, which is the fifth term in the sequence. Hmm, $2 + 3 = 5$.

Now try it with the two biggest fractions generated to date, namely

$$\frac{m}{n}=\frac{7}{5} \quad \text{and} \quad \frac{p}{q}=\frac{41}{29}$$

Take it that $m = 7$, $n = 5$, $p = 41$, and $q = 29$.

This should be interesting. We get

$$\frac{mp+2nq}{mq+np}=\frac{(7\times41)+2(5\times29)}{(7\times29)+(5\times41)}=\frac{577}{408}$$

Look at this! We have skipped over the next two fractions after $\frac{41}{29}$, that is, $\frac{99}{70}$ and $\frac{239}{169}$, to arrive at $\frac{577}{408}$, which is the eighth term in the sequence. Not only are we picking up speed, but in terms of positions it's just $3 + 5 = 8$.

Meaning?

> It seems that when we combine the fraction in position a, say, with the one in position b, say, the process generates the fraction in position $a + b$.

You might be on to something here. Do you realize that you have just used symbols rather than numbers to explain what we'll term your positions conjecture.

> I have become infected! If I could prove this positions conjecture, then I'm sure it'll answer my "doubling conjecture."

The one regarding the positions of the Heron fractions in the fundamental sequence.

> The very one; it has been annoying me.

We've hardly started and you are already making new conjectures and looking for proofs of old ones.

> Are we getting ahead of ourselves?

Goes with the territory. It can be hard to hold a steady pace when investigating. New lines of inquiry seem to spring up all over the place and take one far away from the starting point. Returning to which, you could say that our newly discovered combination rule keeps everything in the family.

> You mean in the infinite family of fundamental fractions?

Yes. I think that this mysterious mathematical dance between $\frac{m}{n}$ and $\frac{p}{q}$ must be awarded top marks for producing such an interesting new rule.

> Ten out of ten, then.

Using the Stars

> I know that each fraction in the fundamental sequence appears only once, but what would happen if we used the same fraction in the combination rule as both $\frac{m}{n}$ and $\frac{p}{q}$?

Now you are being thorough! Why don't you experiment with two fractions that are equal?

> Okay. I'll test the smallest fraction, $\frac{1}{1}$. With $m = p = 1$ and $n = q = 1$ we get

$$\frac{mp + 2nq}{mq + np} = \frac{(1 \times 1) + 2(1 \times 1)}{(1 \times 1) + (1 \times 1)} = \frac{3}{2}$$

> — the second term in the sequence.

If you try with $m = p = 3$ and $n = q = 2$, you get

$$\frac{mp+2nq}{mq+np}=\frac{(3\times 3)+2(2\times 2)}{(3\times 2)+(2\times 3)}=\frac{17}{12}$$

— the fourth fraction in the sequence.

Now that I think about it, there is nothing in the argument you gave that stops us from having $\frac{p}{q}$ and $\frac{m}{n}$ equal.

There is not; everything still holds true. You've answered your own question. We may combine a fraction from the fundamental sequence with itself and still get another fraction from the same sequence.

Well, that wasn't so bad.

Let us explicitly emphasize the fact that $\frac{m}{n}$ combines with $\frac{p}{q}$ to produce the fraction $(mp+2nq)(mq+np)$ by writing the combination rule as

$$\frac{m}{n}*\frac{p}{q}=\frac{mp+2nq}{mq+np}$$

Here the star symbol * stands for the operation that produces a third fraction in the manner indicated.

Not a simple recipe when you're first shown it.

But not much harder than the plus operation for two fractions.

$$\frac{a}{b}+\frac{c}{d}=\frac{ad+bc}{bd}$$

Now that you mention it, I suppose not.

Anyway, from what we have just done, we may write that

$$\frac{1}{1}*\frac{1}{1}=\frac{3}{2}$$

and

$$\frac{3}{2}*\frac{3}{2}=\frac{17}{12}$$

with this notation.

Looks very unusual.

Takes getting used to. If we replace the $\frac{3}{2}$ in the second equation with its "star" equivalent from the first equation, we may rewrite the second equation as

$$\left(\frac{1}{1}*\frac{1}{1}\right)*\left(\frac{1}{1}*\frac{1}{1}\right)=\frac{17}{12}$$

So that there are four $\frac{1}{1}$'s present?

Yes. Now remove the brackets to get

$$\frac{1}{1}*\frac{1}{1}*\frac{1}{1}*\frac{1}{1}=\frac{17}{12}$$

—the *fourth* fraction as a starred combination of *four* copies of the seed fraction $\frac{1}{1}$.

> This is fantastic! Because I think I see how to prove the positions conjecture, as you called it.

Go on.

> In terms of this new star operation, it looks as if the fundamental sequence can be written as

$$\frac{1}{1}, \quad \frac{1}{1}*\frac{1}{1}, \quad \frac{1}{1}*\frac{1}{1}*\frac{1}{1}, \quad \frac{1}{1}*\frac{1}{1}*\frac{1}{1}*\frac{1}{1}, \cdots$$

> We already got the third entry when we showed that

$$\frac{1}{1}*\frac{3}{2}=\frac{7}{5}$$

> which is very convenient.

I'm going to drive you crazy and say that all we showed was that

$$\frac{1}{1}*\left(\frac{1}{1}*\frac{1}{1}\right)=\frac{7}{5}$$

How do you know we can take the brackets off and still get the same result?

> But we did a while back, and you didn't object. Surely this can be done?

It can, but hidden little assumptions all need to be thought about and ironed out.

> I'm not sure I'm be cut out for all this caution.

A good word, and an essential trait for any investigator even if it does seem, at times, to be excessive fussiness.

> May I get on with my argument, however shoddy it may be.

Of course, we'll take it for granted that this is all legitimate.

> So I believe that the fifth fundamental fraction has five $\frac{1}{1}$'s with four stars and so on for the next terms, if you get my meaning.

I do, and you already know what you say is true about the first four fractions. How would you prove it for the fifth term?

> I'd wrap two brackets around the first four $\frac{1}{1}$'s and their three stars and then replace the lot with $\frac{17}{12}$. Then I'd calculate that

$$\frac{17}{12}*\frac{1}{1}=\frac{(17\times 1)+2(12\times 1)}{(17\times 1)+(12\times 1)}=\frac{41}{29}$$

> and have the result I fully expected to get.

Now do it for the general fraction $\frac{m}{n}$ to see if you get the next one.

I'll try, but you'll have to help me with notation, I'm sure.
Okay.

First, I don't know how many $\frac{1}{1}$'s are in the representation of the general fraction $\frac{m}{n}$.

You don't have to. Just show the next fraction has one more $*\frac{1}{1}$.

Right. By the combination rule,

$$\frac{m}{n}*\frac{1}{1}=\frac{(m\times 1)+2(n\times 1)}{(m\times 1)+(n\times 1)}=\frac{m+2n}{m+n}$$

This is the expression for the fraction immediately after $\frac{m}{n}$ in the fundamental sequence.

It is. You are nearly there. When $\frac{m}{n}$ is written out in all its glory in front of the star and the $\frac{1}{1}$ on the left-hand side of this equation, the new fraction is seen to have one more $*\frac{1}{1}$. So I think we can safely say that each fraction of the fundamental sequence may be written in the form

$$\frac{1}{1}*\frac{1}{1}*\cdots*\frac{1}{1}*\frac{1}{1}$$

with the number of $\frac{1}{1}$'s being given by the position of the fraction in the fundamental sequence.

This is a great help. We can now say that if $\frac{m}{n}$ is in position a, and $\frac{p}{q}$ in position b of the fundamental sequence, then $\frac{mp+2nq}{mq+np}$ is in position $a+b$.

Yes, because $\frac{m}{n}$ is in position a, then

$$\frac{m}{n}=\underbrace{\frac{1}{1}*\frac{1}{1}*\cdots*\frac{1}{1}*\frac{1}{1}}_{(a-1)*s}$$

Here the number of stars is one fewer than the number of $\frac{1}{1}$'s in the representation.

This is the kind of thing I wouldn't have known how to handle—the notational part.

It takes a while to get the hang of it. Similarly

$$\frac{p}{q}=\underbrace{\frac{1}{1}*\frac{1}{1}*\cdots*\frac{1}{1}*\frac{1}{1}}_{(b-1)*s}$$

because it is in position b. Then

$$\frac{mp+2nq}{mq+np} = \frac{m}{n} * \frac{p}{q}$$

$$= \left(\underbrace{\frac{1}{1} * \frac{1}{1} * \cdots * \frac{1}{1} * \frac{1}{1}}_{(a-1)*s} \right) * \left(\underbrace{\frac{1}{1} * \frac{1}{1} * \cdots * \frac{1}{1} * \frac{1}{1}}_{(b-1)*s} \right)$$

$$\Rightarrow \frac{mp+2nq}{mq+np} = \underbrace{\frac{1}{1} * \frac{1}{1} * \frac{1}{1} * \cdots * \frac{1}{1} * \frac{1}{1} * \frac{1}{1}}_{(a+b-1)*s}$$

because $(a-1) + 1 + (b-1) = a + b - 1$. The +1 in the middle is counting the star between the two sets of brackets.

I'm star-struck, but I do see how the notation works.

Since the final number of stars is $a + b - 1$, we know that the fraction that results from the combination lives in position $a + b$ of the fundamental sequence.

So that settles the positions conjecture.

It does. Let's savor this result a little more and have some fun.

Count me in. What do you have in mind?

Stepping It Out

The result that

$$\frac{m}{n} * \frac{1}{1} = \frac{m+2n}{m+n}$$

tells us that when the typical fraction $\frac{m}{n}$ of the fundamental sequence is combined with the seed fraction $\frac{1}{1}$, the fraction $\frac{m+2n}{m+n}$ is produced. From what we have just proven, the position number of this fraction is just one more than the position number of the fraction $\frac{m}{n}$.

But we know this already.

Agreed, but suppose we didn't. This result would tell us that the fraction $\frac{m+2n}{m+n}$ is the algebraic form of the fraction coming immediately after $\frac{m}{n}$.

So it would lead us to the one-step rule

$$\frac{m}{n} \rightarrow \frac{m+2n}{m+n}$$

— is that what you are saying?

Exactly. And because

$$\frac{m}{n} * \frac{3}{2} = \frac{3m+4n}{2m+3n}$$

tells us that $\frac{3m+4n}{2m+3n}$ is the form of the fraction two steps on from $\frac{m}{n}$, the two-steps rule is

$$\frac{m}{n} \rightarrow \frac{3m+4n}{2m+3n}$$

— something that we also know independently to be true.

I see what you are driving at. Because $\frac{17}{12}$ is the fourth fraction in the sequence, and because the combination rule tells us that

$$\frac{m}{n} * \frac{17}{12} = \frac{17m+24n}{12m+17n}$$

we can say that

$$\frac{m}{n} \rightarrow \frac{17m+24n}{12m+17n}$$

is the four-step rule. Am I right?

Absolutely. The fraction four steps on from $\frac{m}{n}$ in the fundamental sequence is of the form $\frac{17m+24n}{12m+17n}$. In fact, you can see immediately, by setting $m = 1$ and $n = 1$, that this rule carries $\frac{1}{1}$ into $\frac{17+24}{12+17} = \frac{41}{29}$.

And $\frac{41}{29}$ is the fifth term in the sequence. If the formula is right, it should give me the ninth fraction when I substitute 41 for m and 29 for n. Doing this I get

$$\frac{(17 \times 41)+(24 \times 29)}{(12 \times 41)+(17 \times 29)} = \frac{1393}{985}$$

which is indeed the ninth fraction.

And the next one will be the thirteenth and so on.

This means we now know the general method to construct a rule that will take us any fixed number of steps each time.

We do. If we think of $\frac{m}{n}$ as representing the typical fraction in the fundamental sequence, and of $\frac{p}{q}$ as some fixed fraction, say the one in position r of the fundamental sequence, then the fraction

$$\frac{m}{n} * \frac{p}{q} = \frac{mp+nq}{mq+np}$$

is exactly r steps onward in the sequence from the fraction $\frac{m}{n}$.

> In this case

$$\frac{m}{n} \rightarrow \frac{mp+nq}{mq+np}$$

> is the corresponding r-step rule.

That's it exactly.

> If I want a rule that generates every hundredth term of the fundamental sequence, I work out the hundredth fraction in the fundamental sequence to find the correct p and q to put into the above rule.

Working out the hundredth term with our present knowledge is a bit tiresome, but once you have the correct p and q you are all set and you can start from wherever you like in the sequence.

> And I can do this for any whole number of steps, no matter how large, provided I am willing to find the corresponding p and q?

Yes.

> With this method I can use any speed I desire.

Yes, but it will be constant.

> I realize this. This settles a point we raised a while back. It really does pay dividends to think about things in a fundamental yet simple way, as we have done here. Look at what we know how to do now.

Theoretically, at any rate, we can provide a rule for *any* number of steps because each fraction in the fundamental sequence induces a rule whose speed is given by its location number in the sequence.

Acceleration

> So I know how to travel along the sequence at any speed I like without altering the pressure on the accelerator pedal.

And you now want to be able to vary the speed and accelerate.

> Why not? After all, we are only talking about numbers, so it should be all excitement without any danger.

Since you put it this way, how can I refuse? Well, we already have some experience of accelerating through the fundamental sequence acquired when we were testing our new combination rule. When the first and second terms of the sequence, $\frac{1}{1}$ and $\frac{3}{2}$, are starred, they produce the third term, $\frac{7}{5}$. Then the second term combined with this third term gives the fifth term, $\frac{41}{29}$. When this new term is combined with the previous $\frac{7}{5}$, it gives the eighth term, $\frac{577}{408}$.

> This was when I made the conjecture about the addition of position numbers.

The positions conjecture, which we now know is true.

> So if we continue as we are, the next term will be the thirteenth.

Undoubtedly. Combining the two largest fractions, $\frac{41}{29}$ and $\frac{577}{408}$, obtained to date gives

$$\frac{mp+2nq}{mq+np} = \frac{(41\times577)+2(29\times408)}{(41\times408)+(29\times577)} = \frac{47321}{33961}$$

We could continue in this manner to generate a subsequence of the fundamental sequence, which begins

Fibonacci positions

$$\frac{1}{1}, \ \frac{3}{2}, \ \frac{7}{5}, \ \frac{41}{29}, \ \frac{577}{408}, \ \frac{47321}{33961} \ \ldots$$

and whose terms approach $\sqrt{2}$ at an accelerated rate.

> Because the differences between position numbers are increasing. Can we improve on what we are doing?

In various ways. One such is to introduce a slight modification to what we are doing right now. As things stand, choosing the two most recently generated fractions to generate a fresh one sees the first fraction acting very much as the junior partner because of its magnitude.

> Can we avoid this?

Yes. Just let $\frac{m}{n}$ and $\frac{p}{q}$ both stand for one and the same fraction, the most recently generated one.

> Of course.

We already discussed and implemented the idea of using the same fraction twice in the combination rule. It has the advantage that we work with only one value, the most-up-to-date one, instead of two. Furthermore, the generating rule becomes simpler with only two letters instead of four to confuse us. The fraction propels *itself* forward.

> Very imaginative.

Let's get to it. Since $\frac{m}{n}$ was our original choice for a typical fraction, let us stick with it and set $p = m$ and $q = n$ in

$$\frac{mp+2nq}{mq+np}$$

to get

$$\frac{m}{n} \rightarrow \frac{m^2 + 2n^2}{2mn}$$

as a new rule for generating successive approximations of $\sqrt{2}$.

I don't believe it; this is Hero's rule!

None other.

Obtained, it seems to me, in an entirely different way.

Quite so. Let us remind ourselves of the subsequence of the fundamental sequence that Hero's rule generates when the seed is $\frac{1}{1}$.

It is the Heron sequence:

$$\frac{1}{1}, \ \frac{3}{2}, \ \frac{17}{12}, \ \frac{577}{408}, \ \frac{665857}{470832}, \ldots$$

which is an improvement on the accelerated subsequence given earlier.

Because of the fine-tuning we performed. These fractions occupy positions $1, 2, 4, 8, 16, \ldots$ in the fundamental sequence and so are greater than or equal to the corresponding fractions in our previous subsequence, which occupy the Fibonacci positions $1, 2, 3, 5, 8, 13, \ldots$

Fibonacci positions?

Prepend a 1 to the sequence $1, 2, 3, 5, 8, 13, \ldots$ and you obtain the famous Fibonacci sequence $1, 1, 2, 3, 5, 8, 13, \ldots$ about which so much has been written.

But about which we will say no more so as not to get distracted?

Sadly, yes. However, in light of our new perspective on the Heron rule as

$$\frac{m}{n} \rightarrow \frac{m}{n} * \frac{m}{n} = \frac{m^2 + 2n^2}{2mn}$$

I think the time has come for you to deal with your "doubling conjecture" which you said has been annoying you.

On and off.

I feel confident that you're about to rid yourself of that annoyance.

If the fraction $\frac{m}{n}$ has position a in the fundamental sequence, then its Heron successor

$$\frac{m}{n} * \frac{m}{n} = \frac{m^2 + 2n^2}{2mn}$$

has position $a + a = 2a$ in this same sequence because we simply add position numbers. This, if I'm not mistaken, proves the doubling conjecture.

Well and truly.

I'm surprised by how easy it turned out to be. However, I realize that we have gained a lot of insight since I first made the conjecture.

I think it fair to say that it wasn't clear back then why the position numbers are as they are.

Somehow I don't feel that thrilled about understanding why this doubling is part of the Heron rule. Even though it was killing me that I couldn't find a simple explanation, it was like a puzzle that is both annoying and entertaining at the same time.

They say that nothing kills a problem like a solution.

More Power

I think I can see how to generate more and more powerful rules at will.

And can you put the pedal to the metal, to use your metaphor from the beginning of this discussion?

Not if what I think is true.

Right. Let's hear your idea.

As you said, the Heron rule can be thought of as a fraction operating on itself through the combination rule to propel itself on to the next fraction.

Its next incarnation. Correct.

It's just as if it's squaring itself except that the operation is not ordinary multiplication but this new star operation.

A wonderful operation.

So why not try "cubing"? I reckon we'll step along the fundamental sequence, trebling the previous position number with each new stride.

Well, that would be outstripping Hero's rule in no uncertain terms. You had better elaborate.

My suggested new rule is

$$\frac{m}{n} \rightarrow \frac{m}{n} * \frac{m}{n} * \frac{m}{n}$$

no more, no less.

Looks attractively simple, but how do you work out the "starry expression" appearing to the right of the long right arrow, →?

I write

$$\frac{m}{n} * \frac{m}{n} * \frac{m}{n} = \frac{m}{n} * \left(\frac{m}{n} * \frac{m}{n}\right)$$

$$= \frac{m}{n} * \frac{m^2 + 2n^2}{2mn}$$

$$= \frac{m(m^2 + 2n^2) + 2n(2mn)}{m(2mn) + n(m^2 + 2n^2)}$$

$$= \frac{m^3 + 6mn^2}{3m^2n + 2n^3}$$

$$\frac{m}{n} * \frac{p}{q} = \frac{mp + 2nq}{mq + np}$$

I use the Heron rule to go from the first line to the second, and the general combination rule to go from the second to the third line.

You've been practicing your algebra. We could also say that your suggested new rule is to combine at each stage the typical fraction $\frac{m}{n}$ with its Heron successor to produce a new fraction.

I didn't think of it like this, but yes. And since we know that $\frac{m}{n}$ and its Heron successor are part of the fundamental sequence, we can be sure that the new fraction is also a fundamental fraction.

. . . Which we could also be sure of for other reasons that we need not elaborate.

What's more, it allows us to see very easily why the stride trebles the position number with each new step taken. If the fraction is in position a, its Heron successor is in position $2a$ . . .

. . . and so their starry off-spring is in position $3a$, treble the position number of a. Very convincing. In the above calculation you chose to put the brackets around the last two terms.

I assume and hope that it doesn't matter where I put them.

This is one of those fussy things that we should get the all-clear on before assuming it's true. But we'll take it to be valid so as to get on to the interesting stuff. Your new "cubic rule," as I will dub it, is

$$\frac{m}{n} \rightarrow \frac{m^3 + 6mn^2}{3m^2n + 2n^3}$$

and is an even more powerful rule than Hero's for generating successive approximations of $\sqrt{2}$.

Because it trebles the position number.

When applied to the seed $\frac{1}{1}$, this cubic rule yields the subsequence

$$\frac{1}{1},\ \frac{7}{5},\ \frac{1393}{985},\ \frac{10812186007}{7645370045},\dots$$

of the fundamental sequence.

It is obvious from just these fractions alone that this cubic rule generates terms that approach $\sqrt{2}$ more rapidly than the corresponding terms given by the Heron rule. I wonder how good that fourth fraction is as an approximation of $\sqrt{2}$.

Before you decide, tell me if it is an underestimate or an overestimate.

The fractions shown are the fundamental fractions 1, 3, 9, and 27 and so are all under-estimates of $\sqrt{2}$ because they are in odd-numbered positions.

So the fourth fraction underestimates $\sqrt{2}$ "slightly." We have

$$\frac{10812186007}{7645370045} = 1.414213562373095048795640080754\dots$$

to thirty places of decimals.

Now I'll calculate the fraction after this one using

$$\frac{m+2n}{m+n} = \frac{10812186007+2(7645370045)}{10812186007+7645370045} = \frac{26102926097}{18457556052}$$

This fraction overestimates $\sqrt{2}$. It is

$$1.414213562373095048802726507359\dots$$

to thirty decimal places.

Since $\sqrt{2}$ is smaller than this number but bigger than the previous one we can say that

$$\sqrt{2} = 1.414213562373095048\dots$$

with its eighteen leading digits after the decimal point known forever more.

Just from a knowledge of the fourth fraction in this cubic subsequence of underestimates.

Powerful! If we seed the cubic rule at the second term, $\frac{3}{2}$, in the fundamental sequence, we'll get an equally good—slightly better in fact—subsequence of overestimates of $\sqrt{2}$.

So where to now?

If we "star" the fraction

$$\frac{m^2 + 2n^2}{2mn}$$

which appears in the Heron rule with itself, we get the following rule

$$\frac{m}{n} \rightarrow \frac{m^4 + 12m^2n^2 + 4n^4}{4mn(m^2 + 2n^2)}$$

which generates

$$\frac{1}{1}, \quad \frac{17}{12}, \quad \frac{157258404803291863353217}{111198484434986813793 8112}, \cdots$$

The third term is a monster.

Because of the presence of the fourth powers in this very powerful new rule, we might call it the "quartic" rule and also because the new fraction is

$$\frac{m}{n} * \frac{m}{n} * \frac{m}{n} * \frac{m}{n}$$

— a "starry" multiplication of four like terms.

A good name then.

The decimal expansion of the third fraction just displayed is

1.41421356237309504880168872420969807856967 18753772

Off the record, this agrees with the decimal expansion of $\sqrt{2}$ to forty-seven decimal places.

And this is just the third term! We certainly have got things moving now.

You can say that again. When the typical fraction $\frac{m}{n}$ is starred with the fraction in the quartic rule, we are led to

$$\frac{m}{n} \rightarrow \frac{m^5 + 20m^3n^2 + 20mn^4}{5m^4n + 20m^2n^3 + 4n^5}$$

Applied to the seed $\frac{1}{1}$ this "quintic" rule generates

$$1, \quad \frac{41}{29}, \quad \frac{1855077841}{1311738121}, \quad \frac{35150432379299856878291310769217176444686263 8841}{24855109097042118972946947337281487102909300 2629}, \cdots$$

a subsequence of underestimates to $\sqrt{2}$ because their position numbers in the fundamental sequence are 1, 5, 25, 125, and so on.

That fourth fraction is some whopper.

Isn't it, though? When the typical fraction is starred with the quintic fraction we obtain the "sextic" rule

$$\frac{m}{n} \to \frac{m^6 + 30m^4n^2 + 60m^2n^4 + 8n^6}{6m^5n + 40m^3n^3 + 24mn^5}$$

which is a fairly intimidating recipe but conceptually no more so than the previous ones. The extra power is coming at the price of more and higher-order terms in the rule. Applied to the seed $\frac{1}{1}$ it gives

$$1, \quad \frac{99}{70}, \quad \frac{30122754096401}{21300003689580}$$

followed by

$$\frac{23906612233037460794198505647273994598441535866192765038445737034350984895981070401}{16904527625178060083483488844298922157853960510127056409424438725613140559391177380}$$

as the first four terms of a sequence of approximations that must approach $\sqrt{2}$ at an incredible pace.

> I can well believe it, judging by the enormous fourth fraction. And we can never put the accelerator right to the floor because there is no floor.

Bottomless, because any rule can be improved on by "starring" it with a predecessor. By the way, from a calculation using the last displayed fraction and its successor

1.41421356237309504880168872420969807856967187537694807317 66797379907324784621070388503875343276415727350138462309 12297024924836055850737212644121497099935831413222666 . . .

gives the first 165 digits of the decimal expansion of $\sqrt{2}$!

CHAPTER 5

Odds and Ends

I'm almost disappointed to see the decimal expansion of $\sqrt{2}$ given to so many decimal places because I have a feeling that our discussion about $\sqrt{2}$ is nearly over. Before you finish, is there any more you can say about $\sqrt{2}$ and the sequence

$$\frac{1}{1}, \quad \frac{3}{2}, \quad \frac{7}{5}, \quad \frac{17}{12}, \quad \frac{41}{29}, \quad \frac{99}{70}, \quad \frac{239}{169}, \quad \frac{577}{408}, \ldots$$

without becoming too technical?

There are a few odds and ends which tie in with some of the results we have established, and we might take a light-hearted look at them without proving every detail.

I'd like that.

Best Approximations

Now that we have obtained a decimal expansion of $\sqrt{2}$ that is accurate to more than 160 decimal places, I think we can safely use decimals in calculations concerning $\sqrt{2}$.

Something we have been careful not to do too much of until now?

Yes. The following table shows the decimal approximations of the first twenty multiples of $\sqrt{2}$ rounded to five decimal places of accuracy. I don't think there are any difficulties hidden in the details connected with doing this.

$$\sqrt{2} = 1.41421 = 1 + 0.41421$$
$$2\sqrt{2} = 2.82843 = 3 - 0.17157$$
$$3\sqrt{2} = 4.24264 = 4 + 0.24264$$
$$4\sqrt{2} = 5.65685 = 6 - 0.34315$$

$$5\sqrt{2} = 7.07107 = 7 + 0.07107$$
$$6\sqrt{2} = 8.48528 = 8 + 0.48528$$
$$7\sqrt{2} = 9.89949 = 10 - 0.10051$$
$$8\sqrt{2} = 11.31371 = 11 + 0.31371$$
$$9\sqrt{2} = 12.72792 = 13 - 0.27208$$
$$10\sqrt{2} = 14.14213 = 14 + 0.14213$$
$$11\sqrt{2} = 15.55634 = 16 - 0.44366$$
$$12\sqrt{2} = 16.97056 = 17 - 0.02944$$
$$13\sqrt{2} = 18.38447 = 18 + 0.38447$$
$$14\sqrt{2} = 19.79898 = 20 - 0.20102$$
$$15\sqrt{2} = 21.21320 = 21 + 0.21320$$
$$16\sqrt{2} = 22.62741 = 23 - 0.37259$$
$$17\sqrt{2} = 24.04163 = 24 + 0.04163$$
$$18\sqrt{2} = 25.45584 = 25 + 0.45584$$
$$19\sqrt{2} = 26.87006 = 27 - 0.12994$$
$$20\sqrt{2} = 28.28427 = 28 + 0.28427$$

As you can see, each multiple is also written in terms of the integer closest to it, plus or minus its approximate distance from this integer.

Is there always a closest integer? Might there be some multiple of $\sqrt{2}$ that is *exactly* half-way between two integers?

No. Because this would make the multiple of $\sqrt{2}$ equal to a rational number.

And so make $\sqrt{2}$ a rational number also. I should have seen this for myself. So each multiple has a nearest integer.

Yes; and what I would like to focus on is *how* near, in decimal terms, each multiple is to an integer.

So to begin, $\sqrt{2}$ is at a distance 0.41421 to the right of the integer 1.

This distance is not exact, but a rounded approximation.

I understand.

What about the second multiple?

The number $2\sqrt{2}$ is at a distance 0.17157 to the left of the number 3 on the number line.

An improvement, because 0.17157 is less than 0.41421.

The number $3\sqrt{2}$ is 0.24264 units from the number 4, and to the right of it. It is not as close to 4 as $2\sqrt{2}$ is to the number 3.

So $3\sqrt{2}$ is not as close to its nearest integer as $2\sqrt{2}$ is to its nearest integer.

The next one is even worse. The multiple $4\sqrt{2}$ is at a distance of 0.34315 to the left of the number 6.

Thus, $2\sqrt{2}$ is the current record holder for being closest to an integer.

But that is about to change. The next entry in the table says that $5\sqrt{2}$ is at a distance of 0.07107 from the integer 7.

You are right. A separation of 0.07107 units from the nearest integer is by far the smallest separation so far.

It seems we have a new champion multiple—the number 5.

The multiple of $\sqrt{2}$ closest to an integer so far. Now, if you look down the rightmost column, you won't find any multiple relieving 5 of its title until 12 is reached.

Let me look; absolutely right.

As you can see, $12\sqrt{2}$ is at a distance of 0.02944 from 17, which is less than the 0.07107 units that separate $5\sqrt{2}$ from its nearest integer 7. So 12 wrests the crown of "multiple of $\sqrt{2}$ closest to an integer" from 5.

How long will 12 stay champion?

Until a better multiple is found.

I think at this stage I'm beginning to see what you intend for me to see.

Which is?

That the "record multiples" to date are

$$1, \quad 2, \quad 5, \quad 12$$

and the corresponding "record nearest integers" are

$$1, \quad 3, \quad 7, \quad 17$$

This is exactly what I want you to see. The first set of numbers— the successive champions—are none other than the leading terms of the Pell sequence, while the second set—the corresponding nearest integers—are the first four numbers in what we termed the "first cousin" sequence of the Pell sequence.

This is astonishing!

Alternatively, we can say that the successive fractions in the fundamental sequence

$$\frac{1}{1}, \quad \frac{3}{2}, \quad \frac{7}{5}, \quad \frac{17}{12}, \quad \frac{41}{29}, \quad \frac{99}{70}, \quad \frac{239}{169}, \quad \frac{577}{408}, \ldots$$

have the record multiples for their denominators and the record nearest integers for their numerators.

The number of ways this sequence keeps appearing is almost unbelievable.

Quite so. Now, if we have hit on the true state of affairs, this means that . . .

. . . the next best multiple is 29, with corresponding nearest integer 41.

Yes. Scanning down the right-hand column you'll not see anything smaller than the current minimum separation of 0.02944.

Agreed.

So we have to extend the table to check our prediction. Why don't you do this?

Gladly. The next ten multiples of $\sqrt{2}$ give

$$21\sqrt{2} = 29.69848 = 30 - 0.30152$$
$$22\sqrt{2} = 31.11270 = 31 + 0.11270$$
$$23\sqrt{2} = 32.52691 = 33 - 0.47309$$
$$24\sqrt{2} = 33.94112 = 34 - 0.05888$$
$$25\sqrt{2} = 35.35534 = 35 + 0.35534$$
$$26\sqrt{2} = 36.76955 = 37 - 0.23045$$
$$27\sqrt{2} = 38.18377 = 38 + 0.18377$$
$$28\sqrt{2} = 39.59798 = 40 - 0.40202$$
$$29\sqrt{2} = 41.01219 = 41 + 0.01219$$
$$30\sqrt{2} = 42.42641 = 42 + 0.42641$$

which should be enough.

It is. At the very end of the second to last row we find what we are looking for.

I see it. For the first time since the twelfth row, we have a multiple of $\sqrt{2}$ that is closer than 0.02944 to its nearest integer.

Most welcome, a new minimum separation as predicted. The multiple 29 of $\sqrt{2}$ is within 0.01219 of its nearest integer 41.

I'd check that the next record pair is 77 and 90, but there's a little too much work involved.

Of course, don't dream of doing it.

So it would seem that the successive terms of the Pell sequence

$$1, \quad 2, \quad 5, \quad 12, \quad 29, \quad 70, \quad 169, \quad 408, \ldots$$

provide the successive record multiples of $\sqrt{2}$ that come closest to an integer and that the corresponding record nearest integers are the numbers in the sequence

$$1, \quad 3, \quad 7, \quad 17, \quad 41, \quad 99, \quad 239, \quad 577, \ldots$$

which if true, shows these two sequences in another light.

It is true, but we will not prove it. Instead, let's discuss briefly what further insight it gives us into the fractions in the fundamental sequence

$$\frac{1}{1}, \quad \frac{3}{2}, \quad \frac{7}{5}, \quad \frac{17}{12}, \quad \frac{41}{29}, \quad \frac{99}{70}, \quad \frac{239}{169}, \quad \frac{577}{408}, \ldots$$

— fractions which have been our almost constant companions.

Popping up all over the place.

I want to focus a little more on the fact that of the first twenty-nine multiples of $\sqrt{2}$, the number $29\sqrt{2}$ is the one that comes closest to an integer.

Because, as we have checked, the 0.01219 in

$$29\sqrt{2} = 41 + 0.01219$$

is smaller than the corresponding term for each of the twenty-eight multiples of $\sqrt{2}$ before 29.

The quantity 0.01219 is called the "fractional part" of $29\sqrt{2}$ and represents its separation from its "integer part," which is 41. It doesn't matter, for the purposes of the present discussion, whether it has a plus sign or a minus sign in front of it.

So it's only the size of this fractional part that is important?

Yes. It is the smallest separation observed to date. When we divide the last equation through by 29, we get

$$\sqrt{2} = \frac{41}{29} + \frac{1}{29}(0.01219\ldots)$$

I have added an ellipsis at the end to show that 0.01219 is only an approximation.

All right.

Now I am now going to do the same thing for each of the previous twenty-eight equations appearing in our two tables. That is, I'm going to divide each across by its corresponding multiple of $\sqrt{2}$.

A lot of work.

Perhaps the simplest way to go about this is to first merge our two previous tables but with their second columns of figures deleted. We get a table that shows each of the first thirty multiples of $\sqrt{2}$ in terms of their nearest integer and their fractional parts, as we are now calling them:

$$1\sqrt{2} = 1 + 0.41421$$
$$2\sqrt{2} = 3 - 0.17157$$
$$3\sqrt{2} = 4 + 0.24264$$
$$4\sqrt{2} = 6 - 0.34315$$
$$5\sqrt{2} = 7 + 0.07107$$
$$6\sqrt{2} = 8 + 0.48528$$
$$7\sqrt{2} = 10 - 0.10051$$
$$8\sqrt{2} = 11 + 0.31371$$
$$9\sqrt{2} = 13 - 0.27208$$
$$10\sqrt{2} = 14 + 0.14213$$
$$11\sqrt{2} = 16 - 0.44366$$
$$12\sqrt{2} = 17 - 0.02944$$
$$13\sqrt{2} = 18 + 0.38477$$
$$14\sqrt{2} = 20 - 0.20102$$
$$15\sqrt{2} = 21 + 0.21320$$
$$16\sqrt{2} = 23 - 0.37259$$
$$17\sqrt{2} = 24 + 0.04163$$
$$18\sqrt{2} = 25 + 0.45584$$
$$19\sqrt{2} = 27 - 0.12994$$
$$20\sqrt{2} = 28 + 0.28427$$
$$21\sqrt{2} = 30 - 0.30152$$
$$22\sqrt{2} = 31 + 0.11270$$
$$23\sqrt{2} = 33 - 0.47309$$
$$24\sqrt{2} = 34 - 0.05888$$
$$25\sqrt{2} = 35 + 0.35534$$
$$26\sqrt{2} = 37 - 0.23045$$
$$27\sqrt{2} = 38 + 0.18377$$
$$28\sqrt{2} = 40 - 0.40202$$
$$29\sqrt{2} = 41 + 0.01219$$
$$30\sqrt{2} = 42 + 0.42641$$

We are now ready to divide each row across by the corresponding multiple to get:

$$\sqrt{2} = \frac{1}{1} + \frac{0.41421}{1}$$

$$\sqrt{2} = \frac{3}{2} - \frac{0.17157}{2}$$

$$\sqrt{2} = \frac{4}{3} + \frac{0.24264}{3}$$

$$\sqrt{2} = \frac{6}{4} - \frac{0.34315}{4}$$

$$\sqrt{2} = \frac{7}{5} + \frac{0.07107}{5}$$

$$\sqrt{2} = \frac{8}{6} + \frac{0.48528}{6}$$

$$\sqrt{2} = \frac{10}{7} - \frac{0.10051}{7}$$

$$\sqrt{2} = \frac{11}{8} + \frac{0.31371}{8}$$

$$\sqrt{2} = \frac{13}{9} - \frac{0.27208}{9}$$

$$\sqrt{2} = \frac{14}{10} + \frac{0.14213}{10}$$

$$\sqrt{2} = \frac{16}{11} - \frac{0.44366}{11}$$

$$\sqrt{2} = \frac{17}{12} - \frac{0.02944}{12}$$

$$\sqrt{2} = \frac{18}{13} + \frac{0.38477}{13}$$

$$\sqrt{2} = \frac{20}{14} - \frac{0.20102}{14}$$

$$\sqrt{2} = \frac{21}{15} + \frac{0.21320}{15}$$

$$\sqrt{2} = \frac{23}{16} - \frac{0.37259}{16}$$

$$\sqrt{2} = \frac{24}{17} + \frac{0.04163}{17}$$

$$\sqrt{2} = \frac{25}{18} + \frac{0.45584}{18}$$

$$\sqrt{2} = \frac{27}{19} - \frac{0.12994}{19}$$

$$\sqrt{2} = \frac{28}{20} + \frac{0.28427}{20}$$

$$\sqrt{2} = \frac{30}{21} - \frac{0.30152}{21}$$

$$\sqrt{2} = \frac{31}{22} + \frac{0.11270}{22}$$

$$\sqrt{2} = \frac{33}{23} - \frac{0.47309}{23}$$

$$\sqrt{2} = \frac{34}{24} - \frac{0.05888}{24}$$

$$\sqrt{2} = \frac{35}{25} + \frac{0.35534}{25}$$

$$\sqrt{2} = \frac{37}{26} + \frac{0.23045}{26}$$

$$\sqrt{2} = \frac{38}{27} + \frac{0.18377}{27}$$

$$\sqrt{2} = \frac{40}{28} - \frac{0.40202}{28}$$

$$\sqrt{2} = \frac{41}{29} + \frac{0.01219}{29}$$

$$\sqrt{2} = \frac{42}{30} + \frac{0.42641}{30}$$

In the last column we get a fraction and what I'm going to call the decimal part. These decimal parts are just the fractional parts divided by their corresponding multiples.

> But you didn't bother to work out these decimal parts fully.

Deliberately. You are about to understand why. What I want you to do now is convince me that the decimal part coming after the fraction $\frac{41}{29}$ is smaller in magnitude than any of the other decimal parts appearing in the table.

> This can't be that hard to explain. We already know that 0.01219 is the smallest of all the fractional parts.

That is correct.

> Well, in the table, it is being divided by 29, which is bigger than each of the other divisors 1 to 28.

Correct again, but why is this important?

> It seems obvious to me that the smallest fractional part divided by the biggest multiple is bound to be smaller than all the bigger fractional parts divided by smaller multiples.

And it is. All of which means that $\frac{41}{29}$ is a better approximation of $\sqrt{2}$ than all the fractions shown in the first twenty-eight rows of the table.

It has to be.

Reducing all these fractions to their lowest terms—there are some in need of this, such as $\frac{8}{6}$—and eliminating repetitions such as $\frac{6}{4} = \frac{3}{2}$ gives

$$\frac{1}{1}, \frac{3}{2}, \frac{4}{3}, \frac{7}{5}, \frac{10}{7}, \frac{11}{8}, \frac{13}{9}, \frac{16}{11}, \frac{17}{12}, \frac{18}{13}, \frac{23}{16}, \frac{24}{17}, \frac{25}{18}, \frac{27}{19}, \frac{31}{22}, \frac{33}{23}, \frac{37}{26}, \frac{38}{27}, \frac{41}{29}$$

I've displayed the fundamental fractions in bold.

I see that.

Each of these fractions is an approximation of $\sqrt{2}$.

Is each successive fraction in this sequence better than its predecessor?

No, not by any means. The fraction $\frac{18}{13}$ is not a patch on $\frac{17}{12}$ as an approximation to $\sqrt{2}$, nor is $\frac{23}{16}$, although this latter fraction is an improvement on $\frac{18}{13}$.

And is $\frac{17}{12}$ better than all the fractions between it and $\frac{41}{29}$?

No. The fraction $\frac{24}{17}$ is closer to $\sqrt{2}$ than $\frac{17}{12}$, which is something you might like to show without using decimal approximations.

I'll try it later. But $\frac{41}{29}$ is better than all its predecessors in this list, as you have just said.

It is, but it is no harm to go back over the reason. What you said a moment ago amounts to saying that

$$\frac{1}{29}(0.01219)\cdots < \frac{1}{q}(\text{fractional part}\ldots)$$

whenever q is one of the first twenty-eight natural numbers. This means that $\frac{41}{29}$ is closer to $\sqrt{2}$ than all these other fractions, and so $\frac{41}{29}$ is the best of these in terms of approximating $\sqrt{2}$.

Clearly.

However, $\frac{41}{29}$ is a better approximation to $\sqrt{2}$ than *any* fraction of the form $\frac{p}{q}$, where the denominator q is less than 29 and where the numerator p is any integer.

Not just the numerators shown in the fractions just listed?

Any numerator. The reason is simple: it's because the p corresponding to a given q in the above list is the best numerator for that particular denominator. This is not hard to see if you think about it.

Maybe not for you but I'm going blind from all these fractions. Would you show me for the case of $\frac{16}{11}$, say?

All right. We know from our original table that $11\sqrt{2}$ is within half a unit of 16.

> Because 16 is the nearest integer to $11\sqrt{2}$?

Certainly. Hence the fraction $\frac{16}{11}$ is within half of one-eleventh of $\sqrt{2}$.

> The one-eleventh coming from dividing across by eleven to get the fractional approximation?

As you say. Now, any other fraction of the form $\frac{p}{11}$ is at least one-eleventh from $\frac{16}{11}$.

> The nearest fractions with denominator 11 to $\frac{16}{11}$ are $\frac{15}{11}$ and $\frac{17}{11}$.

Yes. And since $\sqrt{2}$ is within half of an eleventh from $\frac{16}{112}$, it must be more than half an eleventh from these fractions and any other fraction of the form $\frac{p}{11}$.

> And so further from $\sqrt{2}$ than $\frac{16}{11}$. I see it now.

So no other fraction with a denominator less than 29 is closer to $\sqrt{2}$ than $\frac{41}{29}$. For this reason, $\frac{41}{29}$ is said to be a "best" approximation to $\sqrt{2}$.

> And $\frac{17}{12}$ is a best approximation also because no fraction with a denominator less than 12 is closer to $\sqrt{2}$ than it.

Correct.

> And only the fractions in the fundamental sequence have this property?

Yes. The fractions

$$\frac{1}{1}, \quad \frac{3}{2}, \quad \frac{7}{5}, \quad \frac{17}{12}, \quad \frac{41}{29}, \quad \frac{99}{70}, \quad \frac{239}{169}, \quad \frac{577}{408}, \cdots$$

are the *best approximations* to $\sqrt{2}$ in the sense that all fractions with a denominator less than theirs are further from $\sqrt{2}$ than they are.

> So a fraction that is closer to $\sqrt{2}$ than $\frac{99}{70}$, say, must have its denominator greater than 70?

Yes. The first one to achieve this is $\frac{140}{99}$, which is ... I won't say.

Ramanujan and Gauss

I'm now going to pose you four problems based on what we have done, and then I will tell you of a puzzle that will introduce us to two of mathematics' finest number theorists.

> Number theorists—mathematicians who study numbers?

The properties of numbers. "Number theory is the queen of mathematics," is how one of the gentlemen you will soon meet once put it.

Is what we have been doing called "number theory"?

In a sense, yes, and in as elementary a fashion as possible, using no more than simple algebra and without using functions or matrices, to name but two pieces of mathematical machinery that can be tremendously helpful.

> It seems to me that we have achieved a lot with nothing more than algebra.

Certainly, but not perhaps as quickly as we could have. I tried to sail close to the mathematical shore, even if it made our journey longer.

> Well, I am very glad that you did. What are the four problems?

First let me tell you that they can all be solved very simply, so you needn't think, from the sound of them, that they are awfully hard.

> Good to hear. I'll keep that in mind.

The problems concern $\frac{1}{\sqrt{2}}$, the reciprocal of $\sqrt{2}$. The first is to explain why this number is irrational; the second is to write down a sequence of fractions, similar to the fundamental sequence, whose successive terms approach it; the third is to write down its infinite continued fraction expansion; and the fourth is to find the first 160 or so digits in its decimal expansion using the most recently obtained decimal expansion of $\sqrt{2}$.

> These don't sound that simple. Thinking about them should keep me occupied for some time.

The puzzle I want to walk you through is connected with the sequence

$$\frac{1}{1}, \frac{3}{2}, \frac{7}{5}, \frac{17}{12}, \frac{41}{29}, \frac{99}{70}, \frac{239}{169}, \frac{577}{408}, \ldots$$

and will, I'm almost certain, give you some idea of how intimately some human beings know and understand numbers.

> Such as the mathematicians you mentioned. I'm hooked already.

G.H. Hardy, whom I have already mentioned, had a great mathematical colleague in the famous Indian mathematician Srinivasa Ramanujan. At some time in his tragically short life, Ramanujan was given the following puzzle:

> The houses on one side of a street are numbered consecutively beginning with the number 1. Find the number

Srinivasa Ramanujan
(1887–1920)

[See chapter note 1.]

of the house which is such that the sum of the numbers on all of the houses to one side of it is the same as the sum of the numbers on all of the houses to the other side of it.

Ramanujan answered by dictating to his friend Mahalonobis a continued fraction and gave the explanation: "Immediately I heard the problem, it was clear that the solution should obviously be a continued fraction; I then thought, 'Which continued fraction?' and the answer came to my mind." Now what do you think of that?

I cannot say I'm flabbergasted because I don't really understand what it is this man achieved. But it does seem to be a lightning-fast response to a problem that I'm still trying to get my head around.

As would most others be, had they just been given this problem.

So there are all these houses in a row numbered 1, 2, 3, and so on, as far as the last house whose number, I note, we aren't told.

Indeed we are not.

This means then that we have two numbers to find in this puzzle: the number of houses on this one-sided street and the number of the particular house with this special property.

Yes. We can think of ourselves as looking directly at the row of houses

with the lower numbers to our left and the higher ones to our right.

Surely this puzzle works only for certain numbers.

I would think so. Not every street with a given number of houses will possess such a house. However, it's nice to know that there are certain streets with a particular house having this special property.

Yes. I didn't think of this because, I suppose, it could be that such a puzzle might not have any solution.

And I presume when it does have a solution, there is only one possible house number.

I might have known you'd ask a strange question like that. How could there be two different houses? If the sum of all the numbers below the lower-numbered house matched the sum of all the numbers above it, then the sum of all the numbers below the higher-numbered one would be in excess of the sum of all the numbers above it.

I agree. I was just asking as mathematicians are trained to ask, "If a solution exists, is it unique?"

I see.

Would you say the puzzle works for a street with exactly one house?

Strange isn't the word! You mean a street with one house works because there are no houses either to the left or the right of it?

Yes, because you could say that their nonexistent sums are equal. It's just a thought.

I'm assuming the house number itself is not used in the reckoning.

By my reading of the puzzle, it is not. It is the sum of all the numbers to the left of it that must match the sum of all the numbers to the right of it.

It obviously can't work for just two houses.

Because there cannot be houses on both sides of the house in question.

Yes. And it doesn't work for three houses:

because the 1 to the left is not equal to the 3 to the right. I hope there is a solution to this problem small enough to be found by brute-force trial and error.

Let's do a search of streets with a total of, at most, ten houses. If nothing else it will get us thinking about the problem.

Okay. Well, for a street with four houses numbered 1,2,3,4, which house is possible?

It can't be either of the end houses, 1 or 4, and it is not 2. House number 3 just misses out since 1 + 2 is one short of 4.

Try five houses.

Well, 1 and 5 are out immediately, being end houses. The number 2 has only 1 to the left of it, and so cannot match 3 + 4 + 5. House number 3 is out because 1 + 2 is less than 3 + 4, and house number 4 is out because 1 + 2 + 3 is greater than 5.

Another blank. How about six houses?

The end numbers 1 and 6 are out, and there is no point trying 2 or 3 because they're obviously too small. I get the feeling that the house number is going to be near the right end of the street.

The house number is definitely beyond the "middle number or numbers" because it takes a lot of the smaller numbers to balance the larger ones.

House number 4 is ruled out since 1 + 2 + 3 = 6 is not equal to 5 + 6.

While 1 + 2 + 3 + 4 = 10 > 6 rules 5 out as a house number. How about seven houses?

Since $1 + 2 + 3 + 4 = 10 < 6 + 7$, we can rule out all the house numbers up as far as 5; and since $1 + 2 + 3 + 4 + 5 = 15 > 7$, we can rule out house numbers 6 and 7 also.

So, no luck with seven houses. Try eight houses.

This is going to work!

Why?

Because in the case of a street total of seven houses, we saw that $1 + 2 + 3 + 4 + 5 = 15$. This sum matches $7 + 8$.

So?

House number 6 on a street with a total of eight houses:

has

$$1 + 2 + 3 + 4 + 5 = 7 + 8$$

— the sum of the numbers to the left of it matching the sum of the numbers to the right of it.

At last, a solution.

This was hard enough going, though I'm delighted we found one solution.

Before we tackle the puzzle in a professional manner, let me give you a glimpse of some of the other solutions. The next solution is house number 35 in a row of 49 houses. Would you like to check it?

I would, but to do so I have to add up the numbers 1 to 34 and, separately, add the numbers 36 to 49, which should take me some time even with a calculator.

Indeed. You have now encountered another problem or puzzle, this time of a purely mathematical nature which brings us to a much-told story about how another great mathematician solved this newly encountered problem.

It seems to me, from the trial-and-error work we have just done to find the second solution, that if we are to succeed in solving the Ramanujan puzzle completely, then we need to know how this person found such sums in general.

You are absolutely right. But before I relate this mathematical tale, let me tell you now that

$$1 + 2 + 3 + \cdots + 32 + 33 + 34 = 595$$

and that

$$36 + 37 + 38 + \cdots + 47 + 48 + 49 = 595$$

confirming the third solution.

These calculations don't tell me anything about how these sums are found.

That's because I don't want to spoil the story I'm about to recount. When this tale is told, we'll return fully equipped mathematically to solve our house problem.

I'd better let you to it, then.

Carl Gauss

Carl Gauss
(1777–1855)

One of the greatest mathematicians of all time was Carl Friedrich Gauss. It is said that when Gauss was a young boy of no more than eight years of age his teacher asked each member of his class to find the value of the sum

$$1 + 2 + 3 + 4 + \cdots + 97 + 98 + 99 + 100$$

You'll recognize this as similar to, but a little longer than, the first sum we had to do above.

Much longer. This was a hard problem to give to boys of that age.

Which was what the teacher had in mind. He fully expected that this calculation would take each of his pupils most of the class time. As soon as each boy was finished, he was to write his result on his small slate and place it face down in a designated spot in front of the teacher's desk.

They would have needed a lot of time to get all that adding done.

Within seconds, the eight-year-old Gauss placed his slate face down in front of the teacher's desk and went back to his seat, where he remained quietly.

Within seconds? He must have just written down any old number.

The teacher may have thought the very same thing, but if he did, he said nothing. He continued with his reading and left the boy to his thoughts.

And what happened?

As the end of the class period neared, the teacher told the other boys, who were still busy calculating the given sum, to finish up and place their slates in a pile as instructed.

All on top of young Gauss's board?

Yes. When the teacher examined the slates, he found that only one of them had the correct total of 5050.

The one written on young Carl's board?

None other.

Incredible! How did he get this answer so quickly?

Beneath the teacher's long addition he imagined the sum of the numbers from 1 to 100 written in reverse:

$$1 + \ 2 + \ 3 + \ 4 + \cdots + 97 + 98 + 99 + 100$$
$$100 + 99 + 98 + 97 + \cdots + \ 4 + \ 3 + \ 2 + \ 1$$

Now what did he do mentally?

I had better get this right. Did he intend adding each number in the second row to the one above in the top row?

He did.

$$1 + \quad 2 + \quad 3 + \quad 4 + \cdots + \ 97 + \ 98 + \ 99 + 100$$
$$100 + \ 99 + \ 98 + \ 97 + \cdots + \quad 4 + \quad 3 + \quad 2 + \quad 1$$
$$\overline{}$$
$$101 + 101 + 101 + 101 + \cdots + 101 + 101 + 101 + 101$$

Can you see what is so wonderful about this?

He gets a sum of 101 for *each* pair.

Exactly. And can you see how the boy finished off the calculation?

This is terrible pressure. To be put in competition with an eight-year-old! The addition on the final line has one hundred 101s.

It has. Since the overall addition uses each of the numbers 1 to 100 exactly twice, it follows that there are one hundred 101s to be added.

And so

$$2(1 + 2 + 3 + 4 + \cdots + 97 + 98 + 99 + 100) = 100 \times 101$$

giving

$$1 + 2 + 3 + 4 + \cdots + 97 + 98 + 99 + 100 = \frac{100 \times 101}{2} = 5050$$

— the answer Gauss wrote on his chalkboard !

That's how he did it.

It was really very clever how he thought of reversing the numbers and then adding to get the same total each time. It is no wonder he became the great mathematician you said he did, if this is how he was thinking at eight years of age.

They say he could count before he could talk.

    I can well believe it.

This method avoids all the tedium of continually having to add the next number to the sum of all the previous numbers. Because the hundred simple sums $(100 + 1, 99 + 2, \ldots, 1 + 100)$ are really only one sum, Gauss's trick allows him to sidestep all the labour involved in the original task by converting it into a much simpler problem requiring one addition, one multiplication, and one division.

    It is brilliant, and so simple when you see it.

Breathtakingly so, and such economy of effort. The depressing thing is that mere mortals such as we don't see such steps. But, that said, I'm not going to give up singing just because others can do it so much better. By the way, it was Gauss who wrote that, "Mathematics is the queen of the sciences, and number theory is the queen of mathematics."

    I suppose he was entitled to say things like this. May I try out his method on the sums we had above?

Before you do, it might be better for us to obtain a general result that we can use at will.

    Elaborate please.

We use this ingenious trick to sum all the natural numbers up to and including *any* natural number. For example, if I ask you the sum of

$$1 + 2 + 3 + 4 + \cdots + 997 + 998 + 999 + 1000$$

you will no longer gasp with incredulity that you could be expected to work out such a long addition. Instead, imitating the simple pattern of the previous calculations, you will calmly tell me that the answer is

$$\frac{1000 \times 1001}{2} = 500 \times 1001 = 50500$$

Am I not right?

    I'm sure this is exactly what I would do.

If I ask you to explain how you obtained this answer so quickly you might tell me the sum is calculated by the simple rule: *Multiply the largest number by the one after it and halve the result.*

    Of course I'd say this!

In this specific example, the largest number in the addition is 1000, and since the one after it is 1001, we multiply 1000 by 1001 to get 1001000. Then halving this result gives 50500 as the answer.

    As simple as that.

Since "the largest number in the addition" can change from problem to problem, it is useful to use a letter to denote it. Because this largest number is always a natural number, and since n is the initial letter of the word natural, the letter n is often chosen to stand for the largest number in the addition.

So are we going to write down a rule for the sum of the first n natural numbers?

Yes, an all-purpose formula that will give us the flexibility to handle any case. If n is the largest number in the addition, then what is the number just before it?

I suppose it is $n - 1$ with the number before that being $n - 2$.

You are becoming quite comfortable with algebraic notation.

I wouldn't say that, and I have no objection whatsoever when things stay simple with concrete numbers.

We often write the general addition of the first n natural numbers as

$$1 + 2 + 3 + \cdots + (n - 2) + (n - 1) + n$$

with n standing for the largest number in the addition.

So $(n + 1)$ is the number after n, and when this number multiplies n, the result is $n(n + 1)$. Thus, the above rule, when translated, says that the addition sums to

$$\frac{n(n+1)}{2}$$

Yes, pleasingly compact. Thus

$$1+2+3+\cdots+(n-2)+(n-1)+n = \frac{n(n+1)}{2}$$

is the wonderfully simple but powerful general formula giving the sum of the first n natural numbers.

I'm going to recheck the house-street solution of 35, 49.

Very nicely expressed—and brief. Please proceed.

The sum of the first thirty-four house numbers is

$$1+2+3+\cdots+32+33+34 = \frac{34(35)}{2} = 17 \times 35 = 595$$

How am I going to find the sum of the numbers from 36 to 49 inclusive?

A mere technical difficulty, which I have no doubt you'll overcome.

I see how. Find the sum of the first 49 numbers and subtract from it the sum of the first 35 numbers. So

$$36+37+\cdots+48+49=(1+2+\cdots 49)-(1+2+\cdots+35)$$
$$=\frac{49(50)}{2}-\frac{35(36)}{2}$$
$$=(49\times 25)-(18\times 35)$$
$$\Rightarrow 36+37+\cdots+48+49=595$$

as before.

I'm impressed with the simple but clever way you wrote $36 + 37 + \cdots + 48 + 49$ as the difference of two sums, on each of which you could use the general formula.

Thank you.

As a reward you might like to check the next house-street answer is house number 204 on a street with 288 houses.

I'm already on it.

Taking the Bull by the Horns

We are now ready to return to our puzzle and solve it with the help of our newly acquired formula and a little more algebra.

So time to brace myself.

Let us suppose the total number of houses on a street for which this puzzle has a solution is T.

A capital T for the street total, no doubt?

Yes; and let h stand for the elusive corresponding house number, which has house number totals to its left equalling those to its right.

Try describing the nature of the puzzle in terms of the house number h and the street total T.

I'll give it a go. The sum of the numbers up to and including $h - 1$ must be the same as the sum of the numbers from $h + 1$ up to and including T.

Exactly. What is the first sum—the total of all the smaller numbered houses to the left of h?

That would be

$$1+2+\cdots+(h-2)+(h-1)=\frac{(h-1)h}{2}$$

using the general formula with $h-1$ in place of n.

Excellent. What about the second sum—the total of all the house numbers to the right of h?

The second sum is

$$\begin{aligned}(h+1)+(h+2)+\cdots+(T-1)+T &=[1+2+\cdots+(T-1)+T]\\ &\quad-[1+2+\cdots+(h-1)+h]\\ &=\frac{T(T+1)}{2}-\frac{h(h+1)}{2}\end{aligned}$$

using exactly the same idea I used a while back.

First class. Equate these sums, as they say, to see what you end up with.

Setting these sums equal to each other gives

$$\frac{h(h-1)}{2}=\frac{T(T+1)}{2}-\frac{h(h+1)}{2}$$

or

$$h^2-h+h^2+h=T^2+T$$

on multiplying through by 2 and bringing all the h terms to the same side.

Correct. Now the $-h$ and $+h$ terms cancel to give

$$2h^2=T^2+T$$

which is about as simple as we can make it.

Doesn't look too frightening.

We have arrived at a connection between the house number h and the street total T. If we can find an h and a T that fit this equation, then the sums in question will match, and we'll have a solution to the puzzle. Why don't you check the answers we found ourselves?

Are you counting a street with only one house as a solution? Whoever heard of a one-house street?

Well, whether we do or not, see if it fits the equation.

Putting $h=1$ and $T=1$ gives $2(1)^2=1^2+1$, or $2=2$, so it fits the condition.

And our next solution?

Here $h=6$ and $T=8$. Is $2(6)^2=8^2+8$? It is.

Try the other two solutions I told you about just for the fun of it.

> Right.

$$2h^2 = 2(35)^2 = 2450 \quad \text{while} \quad T^2 + T = 49^2 + 49 = 2450$$
$$2h^2 = 2(204)^2 = 83{,}232 \quad \text{while} \quad T^2 + T = 288^2 + 288 = 83{,}232$$

> Both cases check out.

In order to advance, I must perform one or two maneuvers that will seem a little mysterious, but I shall explain the method behind my madness when I'm done.

> Full attention time again.

First, I'm going to multiply both sides of the equation by 4 to get

$$4T^2 + 4T = 8h^2$$

I have interchanged the two sides also.

> I'm with you so far, although I don't know why you did this.

Of course, all will be revealed. Now we add 1 to both sides to get

$$4T^2 + 4T + 1 = 8h^2 + 1$$

Ask yourself if the left-hand side of this expression can be written more compactly.

> If there is factorization of some kind involved, I'm not likely to see it, as it is not a strong point of mine.

Well, the left-hand side can be now written as something squared. We get

$$(2T + 1)^2 = 8h^2 + 1$$

Essentially, $4T^2 + 4T + 1$ has been rewritten as $(2T + 1)^2$—an expression that is the square of the single quantity $2T + 1$. The technique used to arrive here is known as "completing the square."

> You used it already in one of your slick proofs, as you called them.

I did, and it is a very useful idea that shows that our first condition is equivalent to the one just obtained.

> I'm willing to accept this since you treated both sides of the equation equally.

Now viewing $8h^2$ as $2(2h)^2$, we write the last equation in the form

$$(2T + 1)^2 - 2(2h)^2 = 1$$

which brings us very close to previous work.

It does look familiar. It is something squared minus twice something else squared equal to 1.

Exactly. When we replace $2T + 1$ by m, and $2h$ by n, we get

$$m^2 - 2n^2 = 1$$

— an expression you have seen many times.

Now I recognize where you have led us. Every second fraction, beginning with $\frac{3}{2}$, in the fundamental sequence satisfies this relationship, where m is its numerator and n its denominator.

I'm glad you spotted this. Every single fraction in the over-subsequence

$$\frac{3}{2}, \frac{17}{12}, \frac{99}{70}, \frac{577}{408}, \frac{3363}{2378}, \dots$$

satisfies $m^2 - 2n^2 = 1$.

I need to remind myself why we have arrived here.

I can appreciate this, and what you suggest is always the sensible thing to do.

We began by looking for house numbers h and street totals T, which would solve the puzzle, and we have found that any pair of h and T can be obtained from a fraction in the over-subsequence, is that it?

Yes. Specifically, we have shown that a suitable house number is given by half of the denominator of such a fraction.

Because $n = 2h$?

Yes.

I'm getting the idea now. And since $m = 2T + 1$, we can get the corresponding street total by subtracting 1 from the numerator and dividing by 2.

Now you are fully up to speed.

So doing both of these conversions to the numerators and denominators of the over-sequence

$$\frac{3}{2}, \frac{17}{12}, \frac{99}{70}, \frac{577}{408}, \frac{3363}{2378}, \dots$$

gives us the sequence of fractions

$$\frac{1}{1}, \frac{8}{6}, \frac{49}{35}, \frac{288}{204}, \frac{1681}{1189}, \dots$$

which contains all the information about the house numbers and street totals we seek.

Yes, which is very nice. Notice that the fractions in this sequence are not in reduced form.

Indeed they are not.

And for the sake of providing a solution to this puzzle, they shouldn't be brought to lowest form.

Point taken. It is marvelous how this puzzle connects with our $\sqrt{2}$ discussion.

Isn't it? Now we can say, because of all we have established in relation to the fundamental sequence, that the puzzle has an infinite number of solutions.

Powerful.

Mathematics is full of surprises like this. That's part of its great appeal.

So are we to believe that the mathematician Ramanujan saw all of what we have just understood on hearing the puzzle?

You can be sure of it.

Now I really am flabbergasted!

If this is not evidence enough of his genius for numbers, Hardy tells that he once visited Ramanujan in hospital and remarked that the number of the taxi he had taken was 1729, a number that he opined did not strike him as very interesting.

G.H. Hardy
(1877–1947)

I take it then that Hardy was another number theorist.

Reputed to be of the first rank. Ramanujan astounded Hardy by informing him that 1729 is the *smallest* natural number that can be expressed as the sum of two cubes in two different ways, namely as

$$1^3 + 12^3 \quad \text{and} \quad 9^3 + 10^3$$

— as you can easily verify.

I'm speechless!

Different Problem, Same Solution

In solving the last puzzle we made use of the result that

$$1 + 2 + 3 + \cdots + (n-2) + (n-1) + n = \frac{n(n+1)}{2}$$

for each natural number *n*. If we substitute the values 1, 2, 3, ... in turn for *n* in the equation then we get that

$$1 = 1$$
$$1+2 = 3$$
$$1+2+3 = 6$$
$$1+2+3+4 = 10$$
$$1+2+3+4+5 = 15$$

. . . .

The numbers

$$1, \quad 3, \quad 6, \quad 10, \quad 15, \ldots$$

which appear as totals on the right-hand sides of the above equations, are termed the *triangular numbers* for reasons that I hope this diagram makes clear:

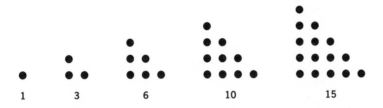

| 1 | 3 | 6 | 10 | 15 |

These look cute.

Alternatively, and more elaborately:

TRIANGULAR NUMBERS	TRIANGLES	SUMS
1 =	●	= 1
3 =	● ●	= 1 + 2
6 =	● ● ●	= 1 + 2 + 3
10 =	● ● ● ●	= 1 + 2 + 3 + 4
15 =	● ● ● ● ●	= 1 + 2 + 3 + 4 + 5 + 6

— to illustrate just the first five.

I can see from the diagrams why these first five numbers are called triangular numbers.

Also, I think they make clear why we'd keep getting triangular arrays by adding extra lines where each has one more black dot than its predecessor.

They do.

Can you see that the first triangle of dots is also a square number, but that none of the other four triangles of dots can be rearranged to form a square?

I can; and this is obvious numerically by observing that none of 3, 6, 10, 15 is a perfect square.

Of course. Now we are going to find all the triangular numbers that are also squares.

So there are others?

Plenty. Let us take a systematic approach and see where it takes us.

Here comes more algebra.

Yes, but we'll be met by a pleasant surprise, which will save us a lot of work.

A surprise? I must watch out for it.

We want to know for what values of n,

$$1 + 2 + 3 + \cdots + (n-2) + (n-1) + n = m^2$$

where m^2 stands for a perfect square. Agreed?

Yes. Can't you replace the left-hand side of this equation by the Gauss formula?

We can and will, to get

$$\frac{n(n+1)}{2} = m^2$$

or

$$n^2 + n = 2m^2$$

Does this equation jog your memory?

Isn't it the same as the equation

$$T^2 + T = 2h^2$$

that we came across in our street problem?

It is, with n instead of the street total T and m instead of the house number h.

Does this mean the two problems are the same?

Well, whether or not the two problems are the same, their answers certainly are.

With the T numbers now being the n numbers and the h answers being the m answers. This is what you meant by a pleasant surprise.

Yes. We know the answers to our current problem because they're the same as those to our previous puzzle.

So the sequence

$$\frac{1}{1}, \quad \frac{8}{6}, \quad \frac{49}{35}, \quad \frac{288}{204}, \quad \frac{1681}{1189}, \ldots$$

contains all the information we need.

The first fraction tells us that the triangular number 1 matches the square number 1 as we already know. Graphically:

● = ●

with the dot on the left representing the first triangular number and the dot on the right representing the first square number. Not earth-shattering from the visual point of view, I know.

The second fraction, $\frac{8}{6}$, tells us that the eighth triangular number is the square $6^2 = 36$.

It does. It says that

$$1 + 2 + 3 + 4 + 5 + 6 + 7 + 8 = 36 = 6 \times 6$$

— a relationship you might like to illustrate geometrically.

That would be fun.

It may take you some time to see how do it.

I think I can show how the triangle can be turned into the square. In this diagram:

I have shown the last two rows of the big triangle differently from its first six rows. These last two rows can be rearranged to form a triangle, as you can see:

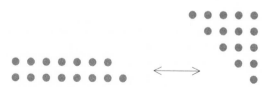

When I place this triangle next to the triangle formed by the first six rows, we can see how the original triangle of eight rows

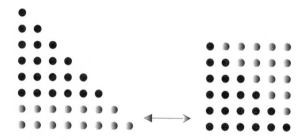

turns into a six-by-six square.

Very nicely done. Your diagram also shows how the six-by-six square can be decomposed to give the large triangle. It is interesting how the last two rows of the overall triangle on the right form a triangle to complement the top triangle with its base of six-by-six dots to give the perfect six-by-six square.

The number of lighter shaded dots in the bottom two rows of the triangle is fifteen, which is the fifth triangular number. This means that the sum of the fifth and the sixth triangular numbers form a six-by-six square.

It does. You might like to check experimentally first, either arithmetically or geometrically, that the sum of some other two consecutive triangular numbers forms a square. Then give a simple algebraic proof of why this is so in general.

I'll try the algebraic proof later, but it seems to me that a geometric proof would be just like the square shown.

You are correct. In this particular case, however, we also have an example of two consecutive triangular numbers forming another triangular number. It is not always true that every pair of consecutive triangular numbers can be combined to give another triangular number.

I understand. In this case, the fifth and sixth triangular numbers add to the eighth triangular number.

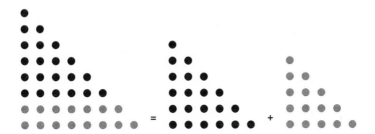

The next solution to our present problem should give us another example.

It should. Let's have all the details.

Right. The third fraction in the sequence

$$\frac{1}{1}, \frac{8}{6}, \frac{49}{35}, \frac{288}{204}, \frac{1681}{1189} \ldots$$

is $\frac{49}{35}$. This result tells us that the forty-ninth triangular number 1225 — the sum

$$1 + 2 + 3 + \ldots + 47 + 48 + 49$$

represented by

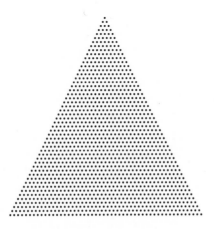

is also the thirty-fifth square, $35^2 = 1225$, represented by

It does. Now this square with its thirty-five rows of thirty-five dots each can be split into two trangles: one of thirty-five rows having dots numbering from 1 to 35 and the other of thirty-four rows having dots numbering from 1 to 34 as you can see from this diagram:

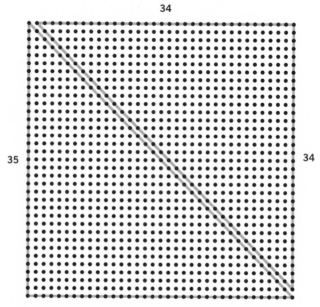

Yes, by slicing the square along one of its diagonals and letting the diagonal dots be part of one triangle. All of which shows that the thirty-fourth and thirty-fifth triangular numbers combine to give the forty-ninth triangular number.

They do. Since the first thirty-five rows of the forty-ninth triangular number can be left as they are to form the thirty-fifth triangular number, it means that rows 36 to 49 inclusively can form the thirty-fourth triangular number.

Couldn't you also leave the first thirty-four rows of the forty-ninth triangular number as they are to form the thirty-fourth triangular number, which would mean that rows 35 to 49 inclusive can form the thirty-fifth triangular number?

This is true also. In terms of the house-street problem, we have

$$1 + 2 + 3 + \cdots + 32 + 33 + 34 = 36 + 37 + 38 + \cdots + 47 + 48 + 49$$

— an equation that tells us that house number 35 on a street with a total of forty-nine houses has the sum of the numbers on the houses below it equal to the sum of the numbers of the houses above it.

As we knew already.

This result shows us that rows 36 to 49 of the forty-ninth triangular number can form the thirty-fourth triangular number.

Because this equation says they have the same sum as 1 to 34.

Now add 35 to both sides of this same equation to get

$$1 + 2 + 3 + \cdots + 32 + 33 + 34 + 35 = 35 + 36 + 37 + 38 + \cdots$$
$$+ 47 + 48 + 49$$

Ah, I see! This new result shows us that rows 35 to 49 of the forty-ninth triangular number can form the thirty-fifth triangular number.

Here is a challenge for you: explain why the numerators in the sequence

$$\frac{1}{1}, \frac{8}{6}, \frac{49}{35}, \frac{288}{204}, \frac{1681}{1189}, \cdots$$

alternate between being perfect squares and twice perfect squares.

I'm glad you told me that they do! Strikes me as a tough nut to have to crack. Maybe I'll think about it after I've finished working on those four problems you already gave me.

Well, only if the mood strikes you.

The Balance of Powers

You mentioned that there were many proofs of the irrationality of $\sqrt{2}$.

I did. Here is a proof by poem:

Written by Maurice
Machover. [See
chapter note 2.]

Double a square is never a square, and here is the reason
why:
If m-squared were equal to two n-squared, then to their
prime factors we'd fly.
But the decomposition that lies on the left has all its
exponents even,
But the power of two on the right must be odd: so one
of the twos is bereaven.

The phrase "double a square is never a square," I recall from my
drill sergeant days, and I know the reason why.

**You do, from our original proof that $\sqrt{2}$ is not expressible as
the ratio of two whole numbers.**

But the three lines that follow the first in this verse are going to
show us why this cannot be the case for different reasons, right?

**Yes, a proof that has to do with the fundamental fact that every
natural number can be expressed uniquely, apart from order,
as the product of prime numbers.**

I see that primes are mentioned in the second line. So this proof
depends on another result, one from arithmetic?

**It uses the result that I just stated and that goes by the name
of the *fundamental theorem of arithmetic*.**

Sounds very important, judging from this title.

**It is, but it is something we almost take for granted without
thinking too much about it. For example, when we write that**

$$6664 = 2 \times 2 \times 2 \times 7 \times 7 \times 17$$

or more briefly as

$$6664 = 2^3 \times 7^2 \times 17$$

**we never entertain the thought that the number 6664 might
have a different "prime decomposition" than the one shown on
the right-hand side of this equation—apart, that is, from jumbling the order of the factors.**

It's hard to imagine how 6664 could be obtained by multiplying different prime numbers by each other.

**Perhaps. The fundamental theorem establishes very carefully
that the decomposition is unique.**

Why is it so fundamental?

We use it all the time even without realising it. For example, if

$$35 = a \times b$$

where a and b are natural numbers greater than 1, what can you say about a and b?

Straight off I'd say that $a = 5$ and $b = 7$, or the other way round.

Completely natural, but you are assuming unique factorization when you draw these conclusions.

If you say so, I suppose I am.

To return to the verse. The start of the second line translates to

$$m^2 = 2n^2$$

while the end of that same line tells us that we should now look at the prime decompositions of the natural numbers m^2 and $2n^2$.

And what are we to make of the third line, "But the decomposition that lies on the left has all its exponents even"?

Well, there is no decomposition displayed in the verse, so to what is he alluding?

Because of what you have just said, I presume he is referring to the prime decomposition of m^2, since it is on the left-hand side of the equation.

This is my reading of it also. Do you understand what is meant by the term *exponent*?

I do. The exponent of the prime 2 in the decomposition of 6664   **$6664 = 2^3 \times 7^2 \times 17$** is 3, while the prime 7 has exponent 2.

Correct, and what is the exponent of the prime 17 in the same decomposition?

I assume it is 1.

It is, even though it is not shown explicitly. We progress.

So what has he in mind when he says, "has all its exponents even"?

Are all the exponents in the prime decomposition of 6664 even?

No. In fact only one of them is even, the exponent of the prime 7.

He is saying that all the exponents in m^2 must be even. Can you see why?

Ah, I think I can see why now. For example, squaring 6664 gives

$$6664^2 = 2^6 \times 7^4 \times 17^2$$

because to square a number in exponent form you simply double the exponent.

Right on. And?

No matter what the exponents are in the prime decomposition of m, and they can be either odd or even, all of the exponents in the prime decomposition of m^2 will be even since twice any natural number is always even.

That's the key point. Now we understand the import of the third line.

So if we can correctly decipher, "But the power of two on the right must be odd: so one of the twos is bereaven," we'll have the proof.

I should think so.

What does he mean by "the power of two on the right must be odd"?

Power is just another word for exponent. He is saying that the exponent of the 2 in the $2n^2$ appearing on the right must be odd. If you can see why this is so, we'll be finished.

Maybe I see what he is driving at now. If the number n has 2 as one of its prime factors, to whatever power, its square, n^2, has 2 to twice that power and so to an even power. For example, if 2^3 is part of the prime decomposition of n, then n^2 has 2^6 in its prime decomposition, right?

Precisely.

Well, then the prime decomposition of $2n^2$ has the prime 2 to an odd power. In my example, 2^7 would be in the prime decomposition of $2n^2$.

Correct. And why would this upset matters?

All the powers in the prime decomposition of m^2 on the left hand-side of the equation

$$m^2 = 2n^2$$

are even, but on right-hand-side the power of 2 in $2n^2$ is odd. So the powers of 2 can't balance.

Excellent. But what if n does not have 2 as a prime factor?

Just as easy. In this case, the prime decomposition of $2n^2$ has 2^1 (which is 2 to the power of the odd number 1) and m^2 has nothing in its prime decomposition to match this odd power.

One way or the other, the exponent of 2 on the right is odd.

As the poet says.

By way of summary: If the prime 2 does not occur in the decomposition of m, then m^2 doesn't have any power of 2 in its prime decomposition. When we replace the m and n in

$$m^2 = 2n^2$$

by their respective decompositions, we end up with no 2s appearing on the left-hand side and at least one 2 on the right-hand side.

Agreed.

But this cannot happen, because $m^2 = 2n^2$ means that m^2 and $2n^2$ are one and the same number. A number cannot have two different prime decompositions according to the fundamental theorem of arithmetic.

And this one has.

It does, because we have just obtained one with no 2 in it and another with at least one 2. "So at least one bereaven two," as the last part of the final line says.

A contradiction is reached if m has no 2 in its decomposition. But if m has 2 to some power in its decomposition, then m^2 has 2 to an even exponent, which cannot be matched by the odd power of 2 on the right.

So, again, some 2 on the right is deprived of a corresponding 2 on the left.

And we arrive at a contradiction again.

Now we understand the poem's proof of the irrationality of $\sqrt{2}$ — a failure to balance the powers of 2—charming.

Infinite Descent

I rather like the proof of the irrationality of $\sqrt{2}$ that I am about to show you now for a reason I hope you'll recognize when it makes its appearance.

Time to be on the watch again.

Suppose that there are natural numbers m and n, say, for which

$$\sqrt{2} = \frac{m}{n}$$

exactly, as we assumed in our first proof.

But we are going to take a different direction this time?

Yes. We begin with an observation we made earlier, which is that $1 < \sqrt{2} < 2$. Replacing $\sqrt{2}$ by the fraction $\frac{m}{n}$, to which it is supposed to be equal, gives

$$1 < \frac{m}{n} < 2$$

An easy step.

Now multiplying this inequality through by the positive integer n gives

$$n < m < 2n$$

Still nice and easy. What's next?

Subtract n from each of the three terms in the inequality to get

$$0 < m - n < n$$

Let me see, $n - n = 0$. Okay, $m - n$ is just itself and $2n - n = n$. I know it's simple but I want to be sure that I get it.

So, in summary, if $\sqrt{2} = \frac{m}{n}$, then

$$1 < \sqrt{2} < 2 \Rightarrow 1 < \frac{m}{n} < 2$$
$$\Rightarrow n < m < 2n \quad (\text{because } n > 0)$$
$$\Rightarrow 0 < m - n < n$$

There are two points here to keep in mind: that m must be a number between n and $2n$, and that $m - n$ is a natural number less than n.

I'll try, but I don't see why we are doing all of this.

Of course, not yet. Now it is also the case that

$$n < m \Rightarrow 2n < 2m \Rightarrow 2n - m < m$$

with $2n - m$ a natural number.

I follow the algebra, even the last step, but let me think why $2n - m$ is a natural number. Might it not be negative?

No, since we have just shown that m is strictly less than $2n$, and so $2n - m$ is positive.

So you have used part of your first argument.

Crucially. Now I want to gather from this what I'm going to need. We have shown that

$$\sqrt{2} = \frac{m}{n} \Rightarrow 2n - m < m \quad \text{and} \quad 0 < m - n < n$$

with $m - n$ and $2n - m$ both positive whole numbers.

It takes a while to get used to these inequalities, but I'll stick with it, as I'm curious to see where all of this is heading.

We'll call on this result after we have established one more. Then we'll be at the end of a preliminary stage.

Just the preliminary stage?

Yes. Now watch very carefully. I'll get you to talk all the steps through afterward.

$$\sqrt{2} = \frac{m}{n} \Rightarrow m^2 = 2n^2$$
$$\Rightarrow m^2 - mn = 2n^2 - mn$$
$$\Rightarrow m(m - n) = n(2n - m)$$
$$\Rightarrow \frac{m}{n} = \frac{2n - m}{m - n}$$

Do you follow any of this?

The first step I've seen before. The second is true because the same quantity, mn, is being subtracted from both sides.

Great so far.

The third step factorizes the previous two sides. In the last step you divide across by quantities that are both positive.

Both positive is vital.

But, again, I don't know why you have taken all of these particular steps.

I know, I have given no motivation for what I'm doing. If you can be patient just a little longer, all will be revealed.

Certainly.

Another summary. Now we may say that

$$\sqrt{2} = \frac{m}{n} \Rightarrow \sqrt{2} = \frac{2n - m}{m - n}$$

What do you make of this?

Nothing, I'm afraid, without having to think a long time about it. Even then I couldn't be sure that I would see what you're hoping I'd see.

Very valid, but you've seen the fraction on the right-hand side before, maybe with different letters.

I'm almost ashamed that I had forgotten. The expression

$$\frac{2n - m}{m - n}$$

is the mechanism we used to go backward along the fundamental sequence from the typical fraction $\frac{m}{n}$ to the one before it.

Given by $\frac{2n-m}{m-n}$, as you say. I was hoping you'd recognize this. Now I'd like you to express the implication

$$\sqrt{2} = \frac{m}{n} \Rightarrow \sqrt{2} = \frac{2n-m}{m-n}$$

in words.

Does it say that if we assume $\sqrt{2}$ is equal to the fraction $\frac{m}{n}$, then $\sqrt{2}$ is also equal to the fraction $\frac{2n-m}{m-n}$?

It does, where m and n are natural numbers. What can you say about the numerator and denominator of the "new" fraction?

That they are also natural numbers?

Yes, and this is very important, but what else can we say? Look back at the result of our preliminary work.

There you showed that

$$\sqrt{2} = \frac{m}{n} \Rightarrow 2n-m < m \quad \text{and} \quad 0 < m-n < n$$

whose significance I think I now see.

Please elaborate.

If I have it correctly, it tells us that the new fraction

$$\frac{2n-m}{m-n}$$

has a smaller numerator than that of $\frac{m}{n}$, since $2n - m < m$, and a smaller denominator than that of $\frac{m}{n}$, since $m - n < n$.

Well spotted, just as it does when $\frac{m}{n}$ is a fraction in the fundamental sequence.

Do we have to worry about the fractions reducing even further?

A good question, to which the answer is no. If the fraction $\frac{m}{n}$ is reduced, which we automatically assume to be the case, then so is the new fraction. You might like to imitate our previous proof, which dealt with reduction.

At the moment I'm quite happy to accept that it's true.

So in

$$\sqrt{2} = \frac{m}{n} = \frac{2n-m}{m-n}$$

the numerator of the second fraction is less than the numerator of the first fraction representing $\sqrt{2}$. Furthermore, the denominator of the second fraction is less than the denominator of the first fraction. And I must add that all these numerators and denominators are natural numbers. What do you make of all this?

> From our experience with moving backward along the fundamental sequence, I suspect that this has to be wrong. Didn't we show that no matter which fraction you start with, the backwards mechanism eventually drops us down to a fraction whose numerator or denominator is no longer positive?

And what would be wrong with that?

> Well, didn't you show very carefully that the fraction obtained using the backward mechanism

$$\frac{2n-m}{m-n} \leftarrow \frac{m}{n}$$

> keeps both the numerator and denominator as natural numbers?

I did. Using "infinite descent," as it is called, we cannot hope to have a positive numerator and denominator at each stage.

> I like the expression "infinite descent."

Because the original m and n are positive integers, this process of continual reduction must fail to produce both a positive numerator and denominator after a finite number of steps.

> Very powerful.

So if you can bear a final summary: when we assume that

$$\sqrt{2} = \frac{m}{n}$$

for some positive integers m and n, we can show that this leads easily to

$$\sqrt{2} = \frac{2n-m}{m-n}$$

with $2n - m$ a positive integer strictly less than m, and $m - n$ a positive integer strictly less than n.

> I wouldn't say "leads easily," but I'm interrupting you.

Then the infinite descent argument scuttles the whole hypothesis by eventually causing some numerator or denominator to become nonpositive, thereby forcing a contradiction.

This is a very nice proof. I probably appreciate it better thanks to having studied how to move backward along the fundamental sequence.

[See chapter note 3.] It is a wonderful idea. The more modern version of the proof sidesteps the infinite descent aspect of this argument by assuming at the outset that if $\sqrt{2}$ can be represented by a fraction $\frac{m}{n}$, then this fraction can be chosen to be in lowest terms.

And is this okay? It does sound reasonable.

Well, it relies on a property of the positive integers known as the "well-ordering principle," which says that every non-empty set of positive integers has a least element.

Which also seems obvious.

Then the accelerated version of the proof uses the implication

$$\sqrt{2} = \frac{m}{n} \Rightarrow \sqrt{2} = \frac{2n - m}{m - n}$$

to arrive at an immediate contradiction, since the second fraction is in lower terms than the assumed lowest form of $\frac{m}{n}$.

Quick and slick, as you might say yourself.

By way of wrapping up this exploration of different proofs of the irrationality of $\sqrt{2}$, I should mention that there are other proofs of the irrationality of $\sqrt{2}$ similar to the ones we have discussed, which can be modified to prove the irrationality of $\sqrt{n}$ for any natural number that is not a perfect square.

So it is easy to prove that

$$\sqrt{2}, \quad \sqrt{3}, \quad \sqrt{5}, \quad \sqrt{6}, \quad \sqrt{7}, \quad \sqrt{8}, \quad \sqrt{10}, \ldots$$

are all irrational numbers as you said earlier?

Yes. If our drill sergeant's original square squadron of soldiers is trebled instead of doubled, the enlarged squadron still cannot be marched in a square formation. Nor can this happen if the squad is increased fivefold or sixfold.

Or by any other multiple that is not a perfect square.

Simply impossible. However, if the original squadron is quadrupled or increased ninefold, or increased n^2-fold for any natural number n, then the enlarged squadron can be paraded in a square formation.

If the squadron is quadrupled, the number of soldiers in each rank and file is doubled.

That's right. If the sergeant wants to increase the length of each rank and file by a natural number n, then the squadron must be increased n^2-fold.

So to treble the rank and file lengths requires that the squadron be increased ninefold. Some consolation for the poor drill sergeant!

Let's hope so, but time for us to march on.

The Four Problems

How did you get on with your four tasks?

It took me some time to get going because I felt I should tackle them in order. But I didn't know how to do the first one at all — to explain in a simple manner why $\frac{1}{\sqrt{2}}$ is an irrational number. I thought I might try to copy the proof of the irrationality of $\sqrt{2}$, but I knew you expected something much easier than this.

Had you some immediate idea how to do some of the others?

Well, I thought I knew an answer to the second one almost immediately. The problem is to write down a sequence of successively improving rational approximations to $\frac{1}{\sqrt{2}}$.

And your solution is?

Because $\frac{1}{\sqrt{2}}$ is the reciprocal of $\sqrt{2}$, I simply turned the fractions in the fundamental sequence

$$\frac{1}{1}, \frac{3}{2}, \frac{7}{5}, \frac{17}{12}, \frac{41}{29}, \frac{99}{70}, \frac{239}{169}, \frac{577}{408}, \ldots$$

upside down to get

$$\frac{1}{1}, \frac{2}{3}, \frac{5}{7}, \frac{12}{17}, \frac{29}{41}, \frac{70}{99}, \frac{169}{239}, \frac{408}{577}, \ldots$$

as one such sequence. It also has seed $\frac{1}{1}$. I must admit that I then took out a calculator and checked, even though I was sure it was right.

So you simply inverted the fractions. Full marks for this solution.

Don't you want a proper proof?

No, the problems are only meant to be fun. Anyway, you were asked to write down a correct sequence, that's all, and I'm more than happy you did this. How did you get on with the continued fraction expansion of $\frac{1}{\sqrt{2}}$?

This scared me at first because continued fraction expansions are new to me—so much so that I was convinced I'd never get it. But then I reminded myself that you had promised the solutions were easy to find. Still, it took a long time before it suddenly dawned on me how simple it is.

How simple is it?

Just form a fraction with the 1 of $\frac{1}{\sqrt{2}}$ in the numerator and write the continued fraction expansion of $\sqrt{2}$ as the denominator to get

$$\frac{1}{\sqrt{2}} = \cfrac{1}{1 + \cfrac{1}{2 + \cfrac{1}{2 + \cfrac{1}{2 + \cfrac{1}{2 + \ddots}}}}}$$

as a solution.

Well done, and full marks again.

I also took a long time to figure out how to get the decimal expansion of $\frac{1}{\sqrt{2}}$ to those 160 or so decimal places that you asked for. My main difficulty was that I didn't quite know how to handle

$$\frac{1}{\sqrt{2}}$$

although we have done many calculations with $\sqrt{2}$.

And what got you moving?

The manipulations you made when discussing the A-series of paper. I remembered that you wrote

$$\frac{2}{\sqrt{2}} = \frac{\sqrt{2} \times \sqrt{2}}{\sqrt{2}} = \sqrt{2}$$

I dropped the term in the middle, turned the other two upside down and made them switch sides to get

$$\frac{1}{\sqrt{2}} = \frac{\sqrt{2}}{2}$$

Suddenly life appeared a lot simpler.

You had gotten over your main stumbling block?

Yes. I now knew that $\frac{1}{\sqrt{2}}$ is just one half of $\sqrt{2}$, which I found a little surprising at first.

On the number line, it is midway between 0 and $\sqrt{2}$.

Once I was able to believe that $\frac{1}{\sqrt{2}}$ is one half of $\sqrt{2}$, I saw how easy it is to get the decimal expansion to all those places of accuracy. Just divide the 165-digit expansion we got for $\sqrt{2}$ by 2 to get

0.70710678118654752440084436210484903928483593768847
40365883398689953662392310535194251937671638207863
67506923115456148512462418027925368606322060748549
9679157066 . . .

In all its splendor. You did division by hand, no doubt?

It took hours!

Three down in great style. Only one to go.

The "talky one."

One where you give reasons rather than work with numbers.

I have to show that $\frac{1}{\sqrt{2}}$ is irrational. I know that $\sqrt{2}$ is irrational, and I know also that it is twice $\frac{1}{\sqrt{2}}$.

True.

I argue by contradiction. If $\frac{1}{\sqrt{2}}$ is not irrational, then it is rational.

Yes, since numbers are either rational or irrational.

And twice a rational number is also a rational number.

Correct.

Which would make $\sqrt{2}$ rational, which we know it isn't. So $\frac{1}{\sqrt{2}}$ cannot be rational, therefore it is irrational.

Excellent. Top marks.

I enjoyed these. However, I wouldn't say they are that easy, even if their solutions are straightforward when you find them.

I agree. You have to think about them the right way, which can take time.

And maybe get a flash of inspiration.

The number $\frac{1}{\sqrt{2}}$ can be thought of as the ratio of the length of a side of a square to the length of its diagonal. In trigonometry, it gives the measure of both the sine and cosine of a 45-degree angle.

So another irrational number that also makes its presence felt.

Rational versus Irrational

[See chapter note 4.]

For our final offering we are going to draw a number of pictures using a definite mathematical scheme.

Good, something visual.

Only when we've learned how the method works will I say what it is all about. Then I'll ask you if one particular picture bears any resemblance to anything to be seen in the world around us.

Another test.

To begin, we designate a specific point to be the center of our picture and take the horizontal line pointing eastward from it as a line of reference. Here's a diagram of what I have in mind:

The small circular dot represents the center, and the line a baseline of reference.

A line of reference for what?

For specifying rotations, as you are about to hear. What I want to do now is place a dot at a distance of one unit from the origin and at a clockwise angle of 45 degrees to this line. The diagram:

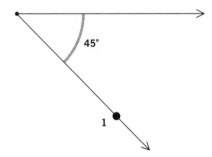

shows this dot and a line from the center through it inclined at an angle of 45 degrees to the horizontal line.

Okay.

When I remove this line passing through the dot and the horizontal reference line, what we should see at this stage is simply

•

•

or on a smaller scale

Not very spectacular — yet, at any rate. Now we'll add another dot inclined at a clockwise angle of 45 degrees to the dot already in place.

   At the same distance from the center?

No. This time we're going to increase the distance to $\sqrt{2}$ units. Here is a diagram with all the scaffolding still in place:

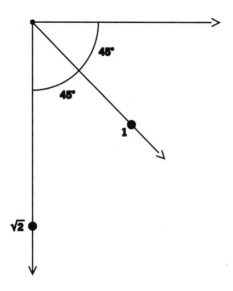

Do you see what I have in mind?

   Yes. So removing the lines and the labels gives, on a smaller scale

which is the picture we are meant to see.

Yes — at this stage. Now we add a third dot inclined at a clockwise angle of 45 degrees to the last dot drawn.

At what distance from the center?

At a distance of $\sqrt{3}$ units from the center point. Here is a full diagram showing all the details so far:

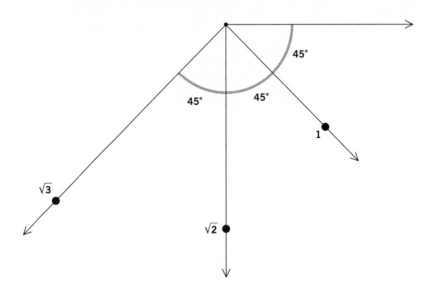

Stripped of everything but what we want to show, it produces

which is still not very revealing.

Do I take it that we continue to add dots in the same way as the ones already?

Yes, with each at a clockwise angle of 45 degrees to its predecessor and at a distance from the center point given by the square root of the dot's number of appearance.

So the distances from the center are getting larger and larger?

Yes. The next dot, which is the fourth, is placed at a distance of $\sqrt{4} = 2$ units from the center and rotated 45-degrees from number-three dot.

And the fifth one is a further 45-degrees from this one at a distance $\sqrt{5}$ from the center.

And so on. Here are the first eight dots placed according to this scheme around the center point:

Can you see the spiral pattern beginning to form?

I can. It's quite easy to see.

Note that the eighth dot is on the baseline since $8 \times 45° = 360°$.

One full revolution. Doesn't this mean that the ninth dot will be in line with the first dot but at a further distance from the center point?

It does. It will be inclined at a clockwise angle of 45 degrees to the baseline and at a distance of $\sqrt{9} = 3$ units from the center. And the tenth dot will be on the same line as the second dot but at $\sqrt{10}$ units from the center.

And the eleventh in line with the third but further from the center, and so on.

Here is a diagram showing how the first sixteen dots are located according to our scheme:

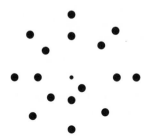

The spiral pattern isn't as pronounced in the outer ring, but if you look at the diagram from a different perspective, you can see arms beginning to form.

Eight arms with two dots on each, is that what you mean?

Yes. Here is the next diagram showing the locations of the first ninety-six dots:

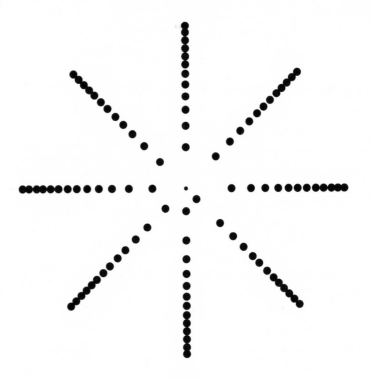

Now the eight arms are clearly visible, while the spiral manner in which the dots take up their positions is only discernible near the center of the picture.

> I can see this. You'd get more arms if you were to use a smaller angle of rotation.

True. If we were to halve our current angle of 45 degrees, we'd end up with sixteen arms like this:

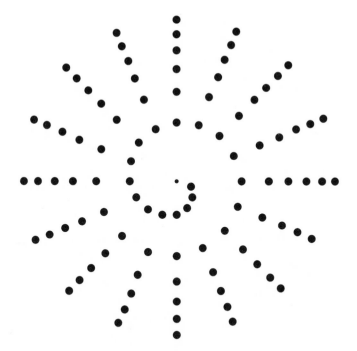

— another pretty picture with its dots spiralling out from the center.

But the spiral is really only prominent for the first revolution.

And is there some way to get a more visible spiral?

Our current angle of rotation equals one-sixteenth, or 0.0625, of a full revolution, and the sixteenth dot positions itself on the baseline. From then on, the subsequent dots start filling out the arms, and it's the arms of the picture that predominate from then on. With an angle of 54 degrees, which is equal to 0.15, or three-twentieths of a full rotation, it takes as far as the twentieth dot to land on the baseline. This happens after three full rotations.

So we get twenty arms, which start filling out from the twenty-first seed onwards.

In the next diagram the first twenty dots are shown shaded in grey.

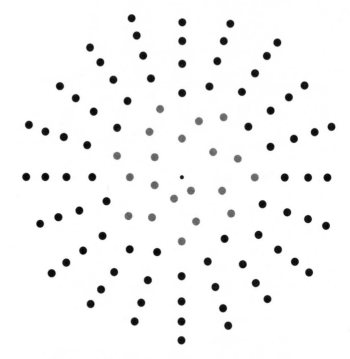

to highlight the three revolutions required to reach the baseline.

Let me just check this. Yes, I see what you are saying.

It's interesting that, by rotating each dot clockwise through 54 degrees from its predecessor, we end up with radial arms separated by 18 degrees. On the other hand, using just twenty rotations through 18 degrees has the twentieth dot landing on the baseline after just one revolution.

Is this better?

Let me answer this question by displaying the first twenty grey dots of the previous diagram alongside the first twenty dots generated by rotating each clockwise through an angle of 18 degrees with respect to its predecessor. Here is the relevant diagram, with the small black dot marking the center point:

What do you notice?

> I notice that only eighteen grey dots are shown, but I guess two are covered by circular black dots.

Yes. The tenth and the twentieth. But how would you compare the distribution of black dots to those of the grey dots?

> I was going to say that a lot more of the black dots are closer to the center than the circles, but this is not true, as I know that, distance-wise, the black dots and the grey dots can be paired, one for one.

Yes, although you might not think so, looking at the diagram. So what might we say, then?

> That the black dots look as if they are arranged better around the center point than the grey dots.

Yes, they use the space better. If the black dots were, say, one arrangement of twenty houses, and the grey dots another on the same overall patch, then I think a group of twenty families would opt for the black-dot distribution.

> I would think so.

Certainly from the point of view of having more individual room and separation from each other. Most would regard the grey dots arrangement a bizarre use of the alloted land.

> I suppose they would.

So by this criterion, a scheme that rotates through 54 degrees is better than one which rotates through 18 degrees.

> This is intriguing. Is there some angle for this scheme that achieves a best possible distribution, whatever that might mean?

I'm told there is, but I won't tell you what it is believed to be until we discuss a little further what we are doing.

> All right.

From what we have done already, we know that some angles are better than others at distributing the same number of dots throughout the available space.

> As is the case with 54 degrees being better than 18 degrees.

Yes. Here the fraction $\frac{3}{20}$ is better than the fraction $\frac{1}{20}$ of a full revolution, but no matter which fraction of a full revolution we choose for our angle of rotation, there will come a stage when some dot lands on the baseline, and from then on the process starts forming arms.

> Is this easy to see?

Yes. If $\frac{p}{q}$, in lowest terms, is the angle of rotation expressed as a fraction of a full revolution, then

$$q \times \frac{p}{q} = p$$

tells us that seed q will fall on the baseline after p full revolutions. Why don't you check out this reasoning on the angle 55 degrees, say?

Right. The angle 55 degrees is the fraction $\frac{55}{360}$ of a full revolution. In lowest terms it is $\frac{11}{72}$, so you are saying that dot 72 will land on the baseline and this will be at the end of 11 full revolutions?

Yes. Check that $11 \times 360 = 3960$ is the same 55×72.

In this, the pattern of dots would have seventy-two arms all radiating from the center with an angle separation of $\frac{360}{72} = 5$ degrees, which is pretty good.

Yes. However, the wedgelike spaces between the radial arms contain no dots. This results in a lot of blank space, particularly further out from the center.

I see what you mean.

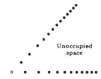

Unoccupied space

A rational multiple of the full angle is not the man for the job of distributing these dots because he leaves vast tracts uninhabited, if I may speak metaphorically.

So from the point of view of good distribution, the rational numbers have this limitation.

That's so.

And can we find a way of populating these blank uniform regions?

Theoretically, at any rate.

How?

By using irrational multiples of the full angle!

Why did you say "theoretically at any rate"?

Because, for practical computation, we can never know an irrational number exactly, so we always end up having to use rational approximations.

Very good ones, I presume.

Yes.

So please explain why theoretically irrational multiples of the full angle are better than rational ones.

For the fundamental but simple reason that two or more dots never get placed on one and the same line emanating from the origin.

I'm going to ask why not, even though I know I should be trying to reason it out for myself.

I'll let you off this time. Suppose the irrational multiple of the full 360-degree angle is r. This means that the angle of rotation between two successive dots is $360r$ degrees. For example, when r is the rational number $\frac{3}{20}$, this angle of rotation is $360(\frac{3}{20}) =$ 54 degrees.

> I understand, but now we need r to be an irrational number, not a fraction.

I know. What would have to happen for two dots to end up being placed on the same line emanating from the origin?

> After the first dot is laid down, the process would have to return to this same line after a finite number of rotations through the given angle of rotation.

Exactly. What does this imply if the number of rotations involved is given by the integer m?

> That $m \times (360r)$ is equal to a certain number of full revolutions.

And if the number of full revolutions is the integer n, what can we say?

> Because a full revolution is 360 degrees, it would mean that
>
> $$m \times (360r) = n(360)$$
>
> But this gives
>
> $$r = \frac{m}{n}$$
>
> which is impossible.

Why?

> Because it would say r is a rational number, which it is not.

Precisely.

> This is fantastic! It's as if the second hand of a clock moved around its face without ever pointing in the same direction more than once.

Exactly. Almost incredible, but it's true. Since this is the case, such rotating is bound to better place the dots in the available space than one that is compelled to cycle through a fixed bunch of rays no matter how numerous they may be.

> I'd have to agree.

I think it is time now that we call on our irrational friend $\sqrt{2}$ to act as a multiplier of the full angle to see what we get.

The $\sqrt{2}$-Flower

> But $\sqrt{2}$ is greater than 1, so surely it is not suitable as a multiplier of the full angle?

Yes and no. Rotating by an angle in excess of 360 degrees is equivalent to rotating in the opposite direction by an angle less than 360 degrees.

> So rotating clockwise through $\sqrt{2}(360) = 509.11688\ldots$ degrees is the same as rotating anti-clockwise through $149.11688\ldots$ degrees?

Yes. And if we want to stick with clockwise rotations, there is nothing stopping us from rotating in this direction through the relevant angle less than 360 degrees.

> Which for the multiple $\sqrt{2}$ means rotating clockwise through $149.11688\ldots$ degrees.

And means that we are simply using $\sqrt{2} - 1$ as the multiplier of the full angle, rather than $\sqrt{2}$.

> This is also an irrational multiplier because $\sqrt{2} - 1$ is irrational.

An irrational multiplier between 0 and 1 that, when applied to one hundred of our dots, produces this picture

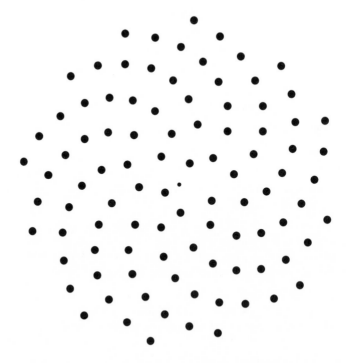

with all its spirals. These are more visible in this next diagram, created by replacing every second dot with a grey dot:

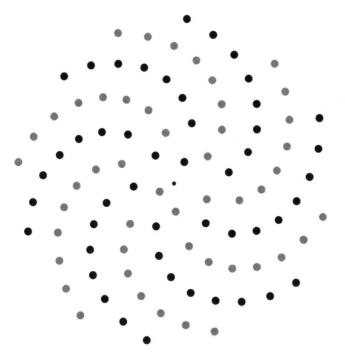

If there are radial arms here, I don't see them.

Nor do I. The dots seem to me to be arranged very nicely. Can we do better?

Before I tell you what I have read about the answer to this question, let me ask if you have ever seen anything in nature that resembles the array of dots just created by the $\sqrt{2} - 1$ multiplier.

So this is the question you had in mind at the outset of this discussion.

Yes.

I was wondering when you were going to ask me to make a connection between nature and the pictures we have been creating mathematically.

And any ideas?

Not really.

True but what if each dot represents a seed or floret that you see on daisies or sunflowers?

[See chapter note 5.]

I'm afraid I have never observed flowers in any great detail.

One model that correctly describes the arrangement of florets in daisies and sunflowers uses the scheme we have described with the distance of a typical dot from the center being scaled by some constant—a distance scaling factor—and a different irrational multiple of the full angle of revolution.

I suppose the distance scaling factor may depend on the flower in question.

Perhaps. The multiple used is one of mathematic's most famous irrational numbers,

$$\frac{1+\sqrt{5}}{2}$$

known as the *golden ratio*.

Golden? The ratio must be very special.

Indeed, but that is a story for another day. Some call the arrangement produced with this golden multiplier the *golden flower*.

An attractive name. So might we not call the arrangement we generated with the multiplier $\sqrt{2}$ the $\sqrt{2}$-flower?

Why not, even if it may not have an actual counterpart in the botanical world. In fact, we could have a virtual garden full of irrational flowers generated by various irrational multipliers.

But the golden one would be the best of all.

So I'm told, in the sense of giving the best possible distribution.

Which is a very good reason.

Since the golden ratio, like $\sqrt{2}$, lies between 1 and 2, we subtract 1 from it and use

$$\frac{\sqrt{5}-1}{2}$$

as the irrational multiple.

According to my calculator, this number's decimal expansion begins 0.61803398 . . . , which when multiplied by 360 gives 222.49223. . . .

So we can choose this number of degrees as our fixed angle of rotation or subtract it from 360 degrees to get 137.50776 . . . degrees and use this as the angle, if we prefer.

And this special angle is going to produce a particularly good arrangement?

Yes. Let me tell you, in terms of continued fractions, which I'm sure you have forgotten all about, that

$$\frac{\sqrt{5}-1}{2}=[0;1,1,1,1\ldots] \quad \text{while} \quad \sqrt{2}-1=[0;2,2,2,2,\ldots]$$

Apparently it is this infinite sequence of 1s and the fact that 1 is the smallest a positive integer can be, that make the golden ratio the best possible multiplier in terms of distribution.

> You are right: I had forgotten our discussion of continued fractions, but I still find this fascinating. So because 2 is the next-smallest positive integer, it may be that the multiplier $\sqrt{2}$ is up with the best of them in terms of being a good multiplier.

Perhaps, I myself would like to learn more about this sometime. Why don't we generate a golden flower with one hundred florets, admire it for a while and afterward superimpose it on our $\sqrt{2}$ floral arrangement to judge visually for ourselves how the two arrangements compare in terms of distribution.

> A good idea.

Here is a golden flower with 100 florets:

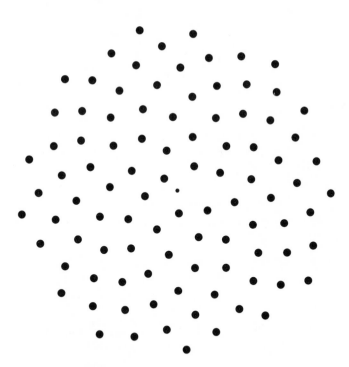

What do you think?

> It looks well, splendid in fact.

Let's see how our $\sqrt{2}$-flower compares with it. Here they are together, the florets of the golden flower in all their splendour beside the $\sqrt{2}$ flower with its grey dots:

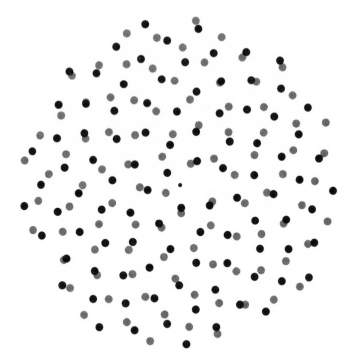

What do you think?

It's hard to say. There doesn't seem to be much difference between them, but perhaps the golden flower is slightly better distributed.

A shade more uniform; visually at any rate. They both look better than the ones derived using rational numbers.

Which for me has been a real revelation.

Why?

This particular discussion has shown me that irrational numbers can be superior to rational ones. I take it for granted that fractions, or rational numbers, are ones that make themselves useful in all sorts of ways.

From cutting cake to musical scales. Whereas $\sqrt{2}$ and other irrational numbers would have to be considered awkward numbers, almost curiosities, that we can really get by without when it comes to most practical tasks?

Perhaps, but the rationals have their own shortcomings, as this distribution of dots problem shows.

This shortcoming is not shared by the irrational community.

I am about to cast off such prejudice, now that I have been further enlightened about the nature of numbers.

Epilogue

The time has come to draw our discussion to a close.

> It has been quite an odyssey and one which I feel has taught me many lessons.

In that case, it has been a journey well worth the undertaking and one which I enjoyed very much.

> As did I, even though when we began I was quite sure that talking about numbers would be of only passing interest to me.

And no one would blame you if that were still the case.

> However, when our little search at the outset to find one square exactly equal to twice another didn't quite pan out as I was hoping it would in fairly short order, I suppose it was then that I became hooked, without my even knowing it.

Ah yes, as someone one wrote, "Mathematics is a trap. If you are once caught in this trap you hardly ever get out again to find your way back to the original state of mind in which you were before you began to investigate mathematics."   [See note 1.]

> Well then, I'm well and truly trapped, because after all the explorations, observations and careful investigations we've been through, I'm sure I've no idea how I used to think.

Alas, you are warped forever!

> I'm afraid so. I'll never be able to think straight again now that I have gained some understanding of how it is mathematicians think, how it is they form conjectures and how they place them on a firm footing afterward.

And this often with no more than a simple use of algebra as we witnessed time and time again.

> But the main thing I've come to realise is that thinking about things for no other reason than just thinking about them for their own sake can be simply great fun.

Chapter Notes

Chapter 1

1. It was Christoff Rudolff who first used the radical sign ($\sqrt{}$) in his 1525 book *Die Coss*. Notice that $\sqrt{}$ is a kind of elongated lower-case "*r*", the first letter in radix, which is Latin for *root*.

   > Ezra Brown "Square Roots from 1; 24, 51, 10 to David Shanks,"
   > *College Mathematics Journal*, Vol 30, No. 2, March 1999, p. 94.

2. A variation of the method appears on page 81 of *The Number Devil* by Hans Enzensberger, Granta Books, London, 2000.

3. Ezra Brown, "Square Roots from 1; 24, 51, 10 to David Shanks," *College Mathematics Journal*, Vol 30, No. 2, March 1999, pp. 83–84.

4. As retold by Choike [2], the discoverer, Hippasus of Metapontum, was on a voyage at the time, and his fellows cast him overboard. A more restrained version by Boyer [1, pp. 71–72] describes both the discovery by Hippasus and his execution by drowning as mere possibilities.

   > D. Kalman, R. Mena, S. Shakriari, "Variations on an Irrational Theme-Geometry, Dynamics, Algebra," *Mathematics Magazine*, Vol 70, No. 2, April 1997. pp. 93–104.
   > 1. Carl Boyer, *A History of Mathematics*, 2nd ed. revised by Uta C. Merzbach, New York, John Wiley & Sons, Inc., 1991.
   > 2. James R. Choike, "The Pentagram and the Discovery of an Irrational number," *College Mathematics Journal* 11, 1980, pp. 312–316.

5. The divisibility proof to be found in Euclid's *Elements* X, (ca. 295 BC), §115a.

Chapter 2

1. Markus Kuhn, *International Standard Paper Sizes*, http://www.cl.cam.ac.uk~mgk25/iso-paper.html, 7/3/02.
2. John Pell (1611–1685) was a great teacher and scholar. Admitted to Trinity College, Cambridge, at the age of thirteen, Pell mastered eight languages before he was twenty. He was professor of mathematics at Amsterdam (1643–1646), and at Breda (1646–1652), and he was Cromwell's representative in Switzerland (1654–1658). He was elected a fellow of the Royal Society in 1663.

   Continued Fractions, by C.D. Olds, New Mathematical Library, New York: Random House Inc., 1963, p. 89.

Chapter 3

1.

$$
\begin{array}{ll}
\begin{array}{r}
m + 2n \\
m + 2n \\
\hline
m^2 + 2mn \\
 2mn + 4n^2 \\
\hline
m^2 + 4mn + 4n^2
\end{array}
& ;
\qquad
\begin{array}{r}
m + n \\
m + n \\
\hline
m^2 + mn \\
 mn + n^2 \\
\hline
m^2 + 2mn + n^2
\end{array}
\end{array}
$$

Chapter 4

1. The Heron method can be regarded as an application of the proposition that the *geometric mean* lies between the *harmonic mean* and the *arithmetic mean*:

$$
\frac{2ab}{a+b} < \sqrt{a.b} < \frac{a+b}{2}
$$

   to the particular case $b = 1$.
2. If $a > \sqrt{2}$ then

$$
\frac{1}{2}\left(a + \frac{2}{a}\right) - \sqrt{2} = \frac{a^2 - 2\sqrt{2}a + 2}{2a}
$$

$$
= \frac{\left(a - \sqrt{2}\right)^2}{2a}
$$

$$
\Rightarrow \frac{1}{2}\left(a + \frac{2}{a}\right) - \sqrt{2} < \frac{\left(a - \sqrt{2}\right)^2}{2\sqrt{2}} \qquad \text{because } a > \sqrt{2}
$$

This inequality show that if the current approximation a satisfies

$$a - \sqrt{2} < \frac{1}{10^d}$$

then the error of the next iteration satisfies

$$\frac{1}{2}\left(a + \frac{2}{a}\right) - \sqrt{2} < \frac{1}{\sqrt{8}} \frac{1}{102^d}$$

That is, the error of the next iteration is less than $\frac{1}{2\sqrt{2}}$ times the square of the error of the current approximation. Because of this, the convergence of the Heron algorithm is quadratic and assures an approximate doubling in the number of decimal places of accuracy with each iteration.

3.

$$
\begin{array}{cc}
m - & \sqrt{2}n \\
p - & \sqrt{2}q \\
\hline
mp - & \sqrt{2}np \\
- & \sqrt{2}mq + 2nq \\
\hline
mp - \sqrt{2}(mq + np) + 2nq
\end{array}
\quad ; \quad
\begin{array}{cc}
m + & \sqrt{2}n \\
p + & \sqrt{2}q \\
\hline
mp + & \sqrt{2}np \\
+ & \sqrt{2}mq + 2nq \\
\hline
mp + \sqrt{2}(mq + np) + 2nq
\end{array}
$$

Chapter 5

1. Ramanajuan's puzzle, *Number Theory with Computer Applications*, taken from R. Kanigel's biography, *The Man Who Knew Infinity: A Life of the Genius Ramanujan*, p. 347.
2. Maurice Machover, St. John's University, Jamaica, NY, 11439, USA and appears in The *Mathematics Magazine*, Vol. 71, No. 2, April 1998, p. 131.
3. David M. Bloom, *A One-Sentence Proof That $\sqrt{2}$ Is Irrational*, from *Mathematics Magazine*, Vol 68, No. 4, 1995, p. 286.
4. Based on the article, *Golden, $\sqrt{2}$, and π Flowers: A Spiral Story* by Michael Naylor, which appeared in *Mathematics Magazine* Vol. 75, No 3, June 2002, pp. 163–172.
5. Due to H. Vogel. See page 100 of *The Algorithmic Beauty of Plants* by P. Prusinkiewicz & A. Lindenmayer, Springer-Verlag, NY, Inc., 1990.

Epilogue

1. Taken from T. W. Körner, *The Pleasures of Counting*, p. viii, who attributes it to E. Colerus, *From Simple Numbers to the Calculus*, Heinemann, London, 1955. English translation from the German.

Acknowledgments

It is a great pleasure to thank all of the following people who helped me in all manner of ways:

Michel Vandyck without whom this tale would never have begun, nor later seen the light of day. He read with great care all of my scribblings as they emerged and remained steadfastly enthusiastic about the whole project from beginning to end.

Donal Hurley, who on reading an earlier draft of the book wrote me a letter full of the warmest encouragement which I will always cherish.

Stephen Webb who refereed the book. In his laudatory but candid report he pointed out a number of significant ways (which he later took great pains to elaborate upon by letter) by which the presentation of the dialogue could be enhanced. This advice was acted upon and has, I believe, resulted in a much improved version of the original. I am much in his debt.

Des Mac Hale, a former and much admired professor of mine, was asked to cast a cold eye over this update. This he duly did with the tremendous energy I had forgotten he possessed. Within weeks I received pages of notes on all aspects of the story along with many new ideas. For his infectious teaching and this most recent wholehearted assistance, *míle buíochas*.

Sarah, our daughter, a thousand thanks also. To her I entrusted the task of reading the book on behalf of all young people with the particular injunction to tell me frankly when she found the presentation boring or unsatisfactory in any way whatsoever. I was assured (as I knew I would be) that she would undertake this responsibility with great zeal and not spare my feelings so as to serve the greater good. Nor was I. Her thorough scrutiny of the manuscript led to greater clarity in the mathematics and the removal of much stuffiness of language.

Elaine, my dear wife, whom I simply cannot thank to any extent that is remotely proportionate to the amount of help she gave me.

Instead I must ask her forgiveness. For no matter how enjoyable a task may be initially, it becomes, by dint of repetition, onerous and dreary. Such was her fate, as I asked her to proof-read successive drafts of pieces I hadn't the talent to get right after one or two attempts. She was my constant companion at every stage of the writing, saved me from numerous spelling errors, solecisms, tedious repetitions and, above all, would not rest until I had made the many different arguments to be found in the narrative as transparent as I possibly could.

Clive Horwood of Praxis Publishing who, believing that the story would appeal, passed it to Copernicus, New York, whom he knew had a special interest in popularizing mathematics. There, Paul Farrell, then Editor-in-Chief, embraced the work with affection; the copy editor Janice Borzendowski added innumerable delicate touches to the text without tampering structurally in the least way with a single line; the text designer and illustration wizard Jordan Rosenblum, along with key assists from Sarah Flannery, helped immeasurably with the technical artwork; David Konopka designed the excellent dust jacket; and Springer senior production editor Michael Koy kept the whole project on track and moving forward, bringing it to the splendid form you find before you.